The Design for Everything Manual

A Guide to Good Design

Henry W. Stoll

HSC Books   Bellevue, WA   2012

ISBN-13: 978-1475231649

ISBN-10: 1475231644

HSC Books
Henry Stoll Consulting
1433 Bellevue Way NE #F, Bellevue, WA 98004

World Wide Web
http://www.designforeverything.info

CONTENTS

PREFACE

The fruits of good design are inherent high quality, low cost, and ease of manufacture and life-cycle support. Over the years, a wealth of design approaches, techniques, short cuts (rules of thumb), and design guidelines have evolved out of design and manufacturing experience that help show the way to good design. Knowledge of this information and the ability to correctly apply it has always been one of the hallmarks of the expert design engineer and manufacturing engineer. My aim in writing this manual has been to capture the essence of this practical knowledge and consolidate it into an easy to understand and apply Design for Everything approach. A manufacturing executive once asked me to make him smart in one minute about a technical subject we were discussing. The goal of this manual is just that, to make the reader smart about good design and how to achieve it in as few words and as quickly as possible.

Design for Everything optimally applies four fundamental rules of good design: the rule of needs, the rule of clarity, the rule of simplicity, and the rule of safety. This is done by using certain cardinal decision principles to guide and inform the critical design choices that define the design. The goal is to eliminate "friction" from the design. Visualize "friction" as all those things such as hassle, uncertainty, and waste that detract from a design to make it less desirable to own and more difficult and expensive to manufacture and support. Good design results when "friction" in the design is reduced to a minimum.

The manual grew out of my many years of teaching a course on Design for "X" in the Master of Product Design and Development Program at Northwestern University, Evanston, Illinois. Design for "X" (DFX for short) is a label applied to a large collection of design methods (e.g., Design for Assembly, Lean Design, etc.) and design guidelines that address particular design issues. The Design for Everything Manual focuses on the principles and practices that underlie the DFX methods rather than on the methods themselves. It covers the same material and addresses the same spectrum of concerns, but in a simpler and more integrated fashion.

The Design for Everything approach is "guided common sense". It is easy to understand and apply regardless of technical background or training. The only prerequisite is a desire to create great designs.

Henry W. Stoll

HOW TO USE THE MANUAL

This manual is about applying the Design for Everything Approach (see below) to engineering design. Section 1 presents the Design for Everything approach and should be read prior to reading the other sections. The remaining sections (2-7) apply the Design for Everything approach to different aspects of design. These sections stand alone and can be read and applied independently. Appendix A provides a list of high level questions useful in preparing for design reviews. Appendix B summarizes several best practices that facilitate the Design for Everything approach at all stages and phases of design.

**The Design for Everything Approach
(the way to good design)**

Good Design is a design in which total design value is maximized.

$$Total\ Design\ Value = \frac{Total\ Quality}{Total\ Cost \times Total\ Time} \rightarrow Maximum$$

Total Design Value is maximized when the Rules of Good Design are optimally applied.

$$Rules\ of\ Good\ Design = \begin{cases} Rule\ of\ Needs \\ Rule\ of\ Clarity \\ Rule\ of\ Simplicity \\ Rule\ of\ Safety \end{cases} \rightarrow Optimally\ Applied$$

Rules of Good Design are optimally applied when Friction is minimized.

$$Friction = \begin{cases} hassel;\ inconvenience; \\ steps;\ process; \\ resistance;\ drag; \\ uncertainty;\ randomness; \\ waste;\ inefficiency \end{cases} \rightarrow Minimize$$

Friction is minimized when all critical design choices are guided and informed by cardinal decision principles.

$$Cardinal\ Decision\ Principles = \begin{cases} Hear\ Voice\ of\ Customer \\ Avoid\ Undesirable\ Interactions \\ Minimize\ Information\ Content \end{cases} \rightarrow Guide\ and\ Inform$$

To Design for Everything, apply the rules of good design, minimize friction, and use the cardinal decision rules to guide and inform all critical design choices.

SECTION 1

GUIDING PRINCIPLES

This section presents the guiding principles upon which the Design for Everything approach is based and attempts to show, though illustrative examples, how these principles can be applied in a disciplined and systematic way to ensure designs that both win in the marketplace and work to maximize the firm's bottom line.

1.1 BASIC CONCEPTS

Purpose: The purpose of this manual is to present an approach to achieving good design that is easy to understand and use by all design decision-makers, regardless of technical background or training. The approach is based on proven design principles and practices that are well-known and widely used by experienced designers. The physical embodiment of a good design can be reverse engineered and its secrets discovered, but the path followed to create the design cannot. Design for Everything teaches the path.

Scope: Although the Design for Everything approach is applicable to all design disciplines, the focus of this manual is on engineering design of products and equipment that are assembled from discrete parts and components.

Good Design: A design in which total design value is maximized,

$$Total\ Design\ Value = \frac{Total\ Quality}{Total\ Cost \times Total\ Time} \rightarrow \max$$

Total Quality: The totality of features and characteristics of a product or device including its design, manufacture, distribution, sale, service, use, and disposal that bears on its ability to satisfy stated or implied needs. Total quality is measured on at least three major dimensions: (1) customer satisfaction with respect to how the design performs its intended function and the experience of owning and using it; (2) robustness against hard to control variation encountered during use, over time, and in manufacture; and (3) ease of manufacture, assembly, and life-cycle support. Good design seeks to maximize total quality.

Total Cost: The sum of all costs, both *direct* and *indirect*, that result from the design, manufacture, distribution, sale, service, use, and disposal of the design over its life-cycle. Direct cost, which includes material, labor, and fixed investment costs, is relatively easy to calculate, and for this reason, it is common practice to base design decisions on direct cost alone. This can be a big mistake, however, because most design decisions impact both direct and indirect cost and indirect costs can far exceed direct costs. Indirect cost includes all the hard to allocate and quantify costs that are typically lumped together and accepted as "overhead" or the "cost of doing business." Good design seeks to minimize total cost.

Total Time: The composite of all times that are affected or determined by design decisions. Reducing the new product introduction cycle or order-to-delivery cycle or product servicing cycle yields numerous benefits. Customer satisfaction is improved, total cost is reduced, and waste, non-value added activity, and quality risk are eliminated. Good design seeks to minimize total time.

1.2 RULES OF GOOD DESIGN[1]

The rules of good design guide all aspects of the Design for Everything approach. Total design value is maximized by applying the rules of good design to each and every design decision in a way the optimally balances all needs and constraints.

Rule of Needs: *Customer needs should be comprehensively understood and then imaginatively satisfied by the design. Designs that follow this rule satisfy user needs better than competitor designs.* This rule dispels the simple belief that the designer knows best what the customer wants and that customers will buy whatever the firm decides to sell. Truly useful and desirable designs result from translating well understood customer needs into designs that satisfy those needs in pleasing and effective ways. The goal is to harmonize the "voice of the customer" with the "voice of the engineer."

Rule of Clarity: *The design should be predictable and unambiguous.* "Clarity" means that the design reliably and consistently performs its function and behaves as intended over time and with use. Product performance and the effect of hard-to-control variation can be accurately predicted. There are no "surprises", hard to explain problems, baffling behaviors, impenetrable uncertainties, manufacturing glitches, or hard to resolve quality issues. When and if engineering change is necessary, changes can be made without undesirable ripple effects and subsequent loss of optimality.

Rule of Simplicity: *The design should be "not complex", "easily understood", and "easily done".* In a simple design, the number of parts, processes, process steps, tools, adjustments, and so forth are a minimum. Parts are shaped so that they are easy to make and assemble. Design simplicity improves reliability and reduces total cost. It helps eliminate waste and hard to control variation, the two main enemies of manufacturing and life-cycle support.

Rule of Safety: *The design should function reliably and safely without harming or endangering humans or the environment.* In a safe design, all types and modes of failure are anticipated, understood, and guarded against in balanced and cost effective ways. Unsafe operating conditions are prevented by design. Humans and the environment are protected from harm, danger, and other hazards. High reliability makes it possible to operate at full capacity without danger or uncertainty. Ideally, the design concept itself inherently avoids and/or eliminates the possibility of danger or harm. When this is not possible, special protective systems are added to ensure safety. Warnings are used when there is no other way of ensuring safety.

---

[1] See Chapter 6 of [1] for further discussion of the rules of clarity, simplicity, and safety. Numbers in brackets refer to references listed in the "Learn More" section at the end of the manual (page 105).

1.3 DESIGN KNOWLEDGE

Design knowledge is the general and specific knowledge that is needed to make high quality design decisions. Good design is not possible unless design knowledge is available when it is needed. Proven best practices that help assure timely availability of accurate design knowledge include: (1) team approach, (2) principle of least commitment, and (3) continuous improvement.

Team Approach: The design project should, from the outset, have continued input from all relevant aspects of the manufacturing enterprise. Experience has shown that this is best achieved using a team approach. In the *team approach*, the design is directed and coordinated by a team of design decision-makers representing all knowledge disciplines essential to the design. Each team member contributes needed information, knowledge, and insight to help inform and improve the quality of design decisions, thereby improving early decisions and avoiding lengthy and costly design iterations.

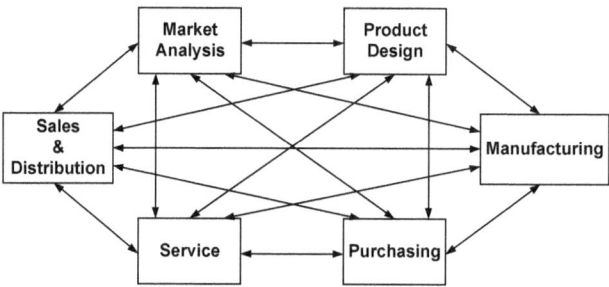

Principle of Least Commitment: When possible, it is best to pursue a policy of least commitment. The principle of least commitment stipulates that *no irreversible decision should be made until it must be made*. By following this principle, decisions that constrain the design direction are postponed thus preserving conceptual maneuverability and making it possible to continue to consider alternative options. A natural extension of this principle is to "design-in" flexibility that allows new technology and/or alternative physical concepts to be adopted when they become available or to be used in future models without having to perform major redesign.

Continuous Improvement: Design changes that further the achievement of good design should be implemented on a continuing basis. Design knowledge increases as the design progresses toward completion. When new or improved design knowledge requires that the design be changed, the best practice of continuous improvement acknowledges that the benefits of good design justify, in the long term, the incremental cost and possible schedule slip required to implement the design change. It also recognizes that incremental changes are usually the most effective. Introducing frequent small improvements is less risky than major redesigns. The key is to design in a way that makes design change easy to implement with a minimum of "ripple effect".

1.4 DESIGN FOR EVERYTHING APPROACH

Experience has shown that the rules of good design are optimally applied to maximize total design value when "friction" is eliminated from the design. *Friction* is the composite of all those things that impede, interfere with, complicate, make difficult, make hard-to-control, make inefficient, or generally detract from achieving good design. The goal of Design for Everything is to reduce friction to a minimum.

$$friction = \begin{cases} hassel;\ inconvenience; \\ steps;\ process; \\ resistance;\ drag; \\ uncertainty;\ randomness; \\ waste;\ inefficiency \end{cases} \rightarrow minimum$$

Cardinal Decision Principles: Design for Everything minimizes friction by relying on three self-evident decision principles to guide and inform each design decision: (1) hear the voice of the customer, (2) avoid undesirable interactions, and (3) minimize information content. Experience has shown that friction at all levels and in all of its forms is minimized by rigorously and systematically applying these decision principles. Think of the cardinal decision principles as guides that optimally apply the rules of good design to filter friction from the design.

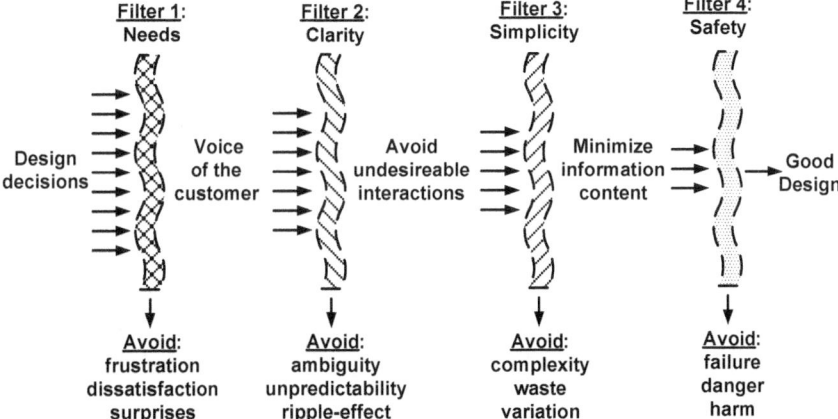

Design Methods: Several structured design methods are included in the Design for Everything approach. These methods are typically simple two or three step procedures that help apply the cardinal decision principles in a disciplined and systematic way. In addition to helping avoid oversight and errors, structured design methods document the decision-making process and help facilitate the team approach by making the process explicit.

1.5 CRITICAL DESIGN CHOICES

Nam Suh [2] defines *design* as the "interplay between <u>what</u> we want to achieve and <u>how</u> we want to achieve it". This interplay involves three critical design choices:

1. **Problem Definition:** <u>What</u> exactly is the design supposed to do? What is its required function?

2. **Physical Concept:** <u>How</u> is the required function to be fulfilled?

3. **Part Decomposition:** <u>How</u> is the physical concept to be divided into separate parts?

Problem Definition

Problem definition is the process of going from a primitive recognition of need to a clear, exact statement of the problem to be solved expressed in engineering terms. The correct problem definition ensures that the right problem is being solved and leads to the "winning" design solution.

Illustrative Example 1.1: Consider the problem faced by a certain TV station. The problem is that ice forms on the antenna tower during inclement winter weather and subsequently falls off, causing harm to people and damage to automobiles below. Concerned about this, the station manager approached a design consultant with the following problem definition: *how to prevent ice from forming on the TV tower*. Although solutions to this problem statement such as installing costly heating elements on the tower structural members could easily be imagined, the consultant, experienced in the need for accurate problem definition, asked the following questions:

*   What would happen if ice did form?
*   What harm would such formation do?

Following this line of reasoning, the design consultant formulated a much broader problem definition: *how to prevent ice that forms on the TV tower from doing harm or damage to people and equipment below the tower*. This second problem definition led to an obvious and cost-effective solution that was quickly accepted: build a shed-like structure to protect against falling ice. Not only was this selected solution a clearly superior solution functionally, it also avoided the costly and time-consuming design and development effort that would have been required by solutions to the first, less accurate problem definition.

Functional Requirements: An important aspect of problem definition is that of defining the functional requirements of the design. *Functional requirements* (or FR's for short) are concise statements of what the design must achieve. "Prevent ice that forms on the TV tower from doing harm or damage to people and equipment below the tower" is the main functional requirement of Illustrative Example 1.1. Some FR's for the design of a plastic beverage container might be "contain axial and radial pressure", "withstand moderate impact when dropped", "allow stacking on top of each other", and "minimize the use of plastic". FR's may be accompanied with specifications of physical quantities and ranges (e.g., radial pressure in the range of 1 to 3 psi). FR's can be classified as main, critical, and auxiliary. *Main* FR's pertain to the overall task to be performed. *Critical* FR's are a subset of the main FR's that are considered to be of critical importance to the design and/or that offer the most opportunity for innovation. *Auxiliary* functional requirements, on the other hand, have a supportive or complementary character and generally arise because of the particular physical concept or part decomposition that is selected.

Functional Decomposition: FR's can be determined by decomposing the problem of design into simpler subproblems. To perform a functional decomposition, first represent the overall function of the design as a single black box operating on material flow (double solid line), energy flow (single solid line), and information flow (dashed line). Divide the single black box into subfunctions to create a more specific description of what the elements of the design must do to implement the overall function. Each subfunction is an FR. Experience has shown that it is usually best to subdivide the overall function into 3 to 10 subfunctions.

Illustrative Example 1.2: Decompose the problem of designing a barbeque grill.

Design Problem:

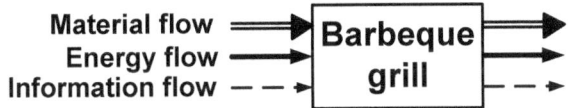

Functional Decomposition:

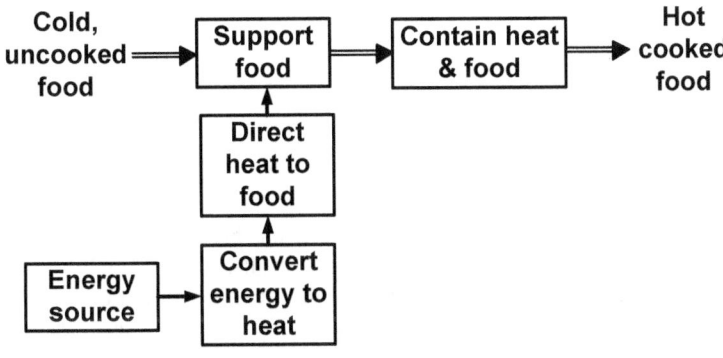

Physical Concept

The *physical concept* includes (1) the physical principles by which the design achieves its overall function and (2) a physical description of the design embodiment. Put simply, the physical concept is how the design works. Developing the physical concept is a huge step. It requires that the design team enter the real world of physical principles, materials, how things work, and how things work together. The choice of physical concept adds a tremendous amount of information to the design, and there is not again in the design process an opportunity for such significant choices. Changing the physical concept once the design has been released is usually extremely difficult and therefore, unless intentionally designed for change (principle of least commitment), the opportunity for selecting the physical concept seldom occurs more than once. This is one of the reasons why "doing it right the first time" is so important.

When all else is equal, the physical concept having the lowest "intrinsic cost" should be selected. *Intrinsic cost* is the theoretically lowest possible total cost for a given physical concept. To achieve the intrinsic cost, the physical concept must be perfectly designed using a theoretical minimum number of ideally shaped parts that allow optimal manufacture and assembly. Intrinsic cost can sometimes be approached over time through a process of continuous improvement.

Illustrative Example 1.3: Consider the design of an air compressor. Several alternative physical concepts are possible. For example, the overall function of raising air pressure could be accomplished by using a reciprocating piston, a scroll, or possibly a rotary vane[2]. Which physical concept should be chosen? Both the scroll and rotary vane compressors require fewer theoretical parts (see Section 4) compared to the reciprocating piston design. And, compared to the scroll, the vane compressor is composed of parts that are easier to manufacture and have less critical tolerances. Therefore, assuming pressure and flow requirements are satisfied equally, the vane compressor is likely to have the lowest intrinsic cost thus providing either greater profit or enhanced price flexibility.

---

[2] A reciprocating piston air compressor works like a car engine. It has a crankshaft and compresses air that is trapped between the piston and the cylinder head. A scroll air compressor uses two interleaving scrolls, each shaped in the form of an involute curve or Archimedean spiral, to compress the pocket of air that is trapped between them. Rotary vane compressors use a vane that reciprocates radially in an eccentrically mounted rotor to compress air that is trapped between the vane, rotor, and housing.

Part Decomposition

Creating the definitive final design (physical embodiment) involves a complex interplay between the physical concept, geometric layout, component configuration, selection between standard and designed components, material selection, manufacturing process selection, and assembly process. Inherent within this interplay are choices about how to divide the design into parts. We call the result of these choices the *part decomposition*. To understand the part decomposition of a design, disassemble a physical example (or imagine how it would be done), keeping track of each part as it is removed, and then reassemble the parts in the reverse order. The number of parts, their geometric arrangement, and the inter-relationships and interfaces between them define the part decomposition.

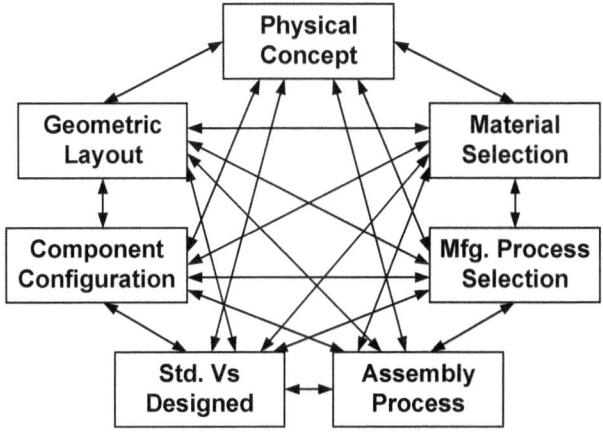

Some designs, such as the air compressor example (Illustrative example 1.3) discussed previously, are "tightly coupled". In these cases, the physical concept dictates the part decomposition. In other designs, such as the design of an electrical junction box, coupling is weak and many alternative part decomposition choices are possible. Most designs fall somewhere between these extremes.

As a general rule, *the best part decomposition is the one having the fewest easy to make and assemble parts*. This is not always the case. For example, it may sometimes be preferable to use separate fasteners to avoid long tooling lead times.

For many designs, the part decomposition choice can be as important as or even more important than the physical concept. The part decomposition determines the ease of assembly and life-cycle support. It also determines the number and complexity of the designed parts, which in turn, influences tooling cost, tolerance stack-up, smoothness of force-flow, and numerous other considerations. As a result, part decomposition has an enormous and far-reaching effect on total design value. Luckily, the part decomposition can usually be evolved and optimized through continuous improvement and cost reduction programs.

1.6 CARDINAL DECISION PRINCIPLES

The premise of Design for Everything is very simple: to achieve good design, optimally apply the rules of good design to maximize total design value by eliminating friction at all levels and in all of its forms. Friction is avoided and minimized when design decisions are guided and informed by three *cardinal decision principles*: (1) hear the voice of the customer, (2) avoid undesirable interactions, and (3) minimize information content. To hear the *voice of the customer* means that the design is based on real customer needs and that it is validated by real customer feedback. *Undesirable interactions* occur when different functions, or aspects, or features of the design act on each other in ways that complicate, confuse, interfere, impede, make uncertain, or otherwise degrade performance, functionality, and/or ease of manufacture. To *minimize information content* means to eliminate and minimize hassle, inconvenience, resistance, uncertainty, and waste from all aspects of the design.

Hear the Voice of the Customer

Customers include all who come in contact with the design. The end customer, or user, is typically given the highest importance rating, but not always. A customer need is any desired quality, attribute, or trait of a potential design that is expressed by actual customers. The voice of the customer is heard by developing a comprehensive understanding of needs through direct interaction with real customers and the use environment, basing the problem definition on customer needs, and validating the design with customer feedback. In addition to being key to good design, this decision rule is an important stimulus for creativity and design innovation. It is discussed at length in Section 2.

Avoid Undesirable Interactions

Experience has shown that many quality, manufacturing, and performance problems can be traced to undesirable interactions between various aspects or functions of a design. Undesirable interactions prevent good design by interfering with functionality and generating "friction". Interactions are mutual or reciprocal actions or influences between different aspects of the design. They become undesirable when functionality is degraded, interfered with, made unreliable, or made unpredictable; when capacity is diminished; when ease of manufacturability is compromised; and when complexity, uncertainty, inefficiency, and waste are increased. For example, the steering function and driving function of a front wheel drive automobile interact undesirably when "torque steer" causes the vehicle to swerve to the left or right during rapid acceleration. Excessive or difficult to solve quality, manufacturing, and performance problems are "red flags" that signal the possible presence of one or more undesirable interactions. Avoiding undesirable interactions requires constant questioning and evaluation of each design decision. How could an undesirable interaction occur? What would be the symptoms? "What-if" this condition arose, what would happen?

There are many ways in which undesirable interactions occur. For example, undesirable interactions can arise due to unwanted or unanticipated deformation caused by force and/or temperature change (see Section 3). Product variety can be another source of undesirable interaction. Changing from one supplier to another can require tooling and process changes if the dimensions or geometry of the supplied component changes. Similarly, different product models sometimes cannot share parts because components such as pumps and electric motors are a different size. Often, an adaptor plate or other interfacing element will decouple these undesirable interactions.

Functional requirements are "coupled" when a change to one FR causes a hard to predict or undesirable change or behavior in another. The most effective way to avoid undesirable interactions such as these is to design so that each functional requirement is fulfilled in a way that is independent of the other functional requirements. This self-evident truth is formalized by Nam Suh [2] as the *independence principle*:

> ### *Maintain independence of functional requirement*

To maintain independence means that each functional requirement (FR) is fulfilled by a separate aspect, feature, or component within the design. Maintaining the independence of functional requirements yields many desirable benefits:

- Each FR is predictable and unambiguous.

- The design is robust[3] against hard-to-control variation, unanticipated surprises, and other disruptive effects.

- The "ripple effect" that often occurs due to "engineering change" is short circuited. This helps prevent sub-optimal design that would otherwise result from engineering change made late in the project.

- The design is simple because interactions are not a consideration.

- Safety and reliability are enhanced because catastrophic failure and/or gradual degradation of the design's capability with respect to any one particular functional requirement does not ripple to or effect capability with respect to other functional requirements.

As convincingly argued by Suh [2], when the independence of the minimum orthogonal set of FR's is maintained, all undesirable interactions will be avoided. This underscores the importance of defining FR's carefully and in such a way that they are independent of each other. As illustrated by the following example, the independence principle presents an important, easy to follow path to good design.

---

[3] Throughout this manual, the term "robust" means that the design is insensitive to hard-to-control variation such as changes in operating conditions, changes due to degradation and deterioration over time and with use, and differences in products manufactured on the same production line and/or under the same specification.

Illustrative Example 1.4: Which radar antenna-mount design is preferred?

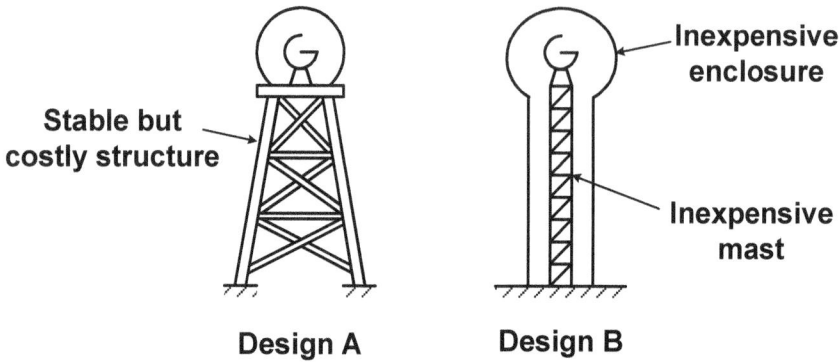

Stable but costly structure

Inexpensive enclosure

Inexpensive mast

Design A   **Design B**

Main Functional Requirements:
    FR1: Support the antenna.
    FR2: Prevent movement due to external disturbances such as wind.

Design A: Satisfies both FR's by using a rigid tower to support the antenna and protect it from unacceptable movement due to wind forces. This design lacks clarity, however, because the functional requirements are "coupled" making the tower behavior with respect to each FR ambiguous and hard to predict. As a consequence, the tower must be carefully designed and constructed using engineering analysis and skilled construction personnel. In addition, the design is not robust against ground induced disturbance such as that caused by a passing freight train because isolating the support structure from the ground will affect the tower's behavior in the wind. This undesirable interaction defies an easy fix.

Design B: Satisfies FR's using a mast to support (FR1) and a barn-like enclosure to protect from wind disturbances (FR2). The independence of the functional requirements is maintained by satisfying each FR using a separate aspect of the design. This provides clarity because behavior with respect to each FR is inherently predictable and unambiguous. The mast is easily designed or better yet, purchased from a supplier. The enclosure can be built using readily available and inexpensive materials by local non-skilled labor. A simple vibration isolation mount can be used to isolate the support mast from ground induced vibration without affecting the way the enclosure behaves in the wind.

The Choice: Both design alternatives satisfy the FR's, but **_Design B_** is preferred because it has inherent clarity, is robust against hard to control variation and engineering change, has a minimum of complexity making it easy to design and construct, and will function reliably and safely because, no matter how severe the wind, failure of the enclosure will not cause the radar antenna mast to fail and failure of the mast will not cause the enclosure to fail.

Logical building block method: This method offers an effective way to systematically generate numerous alternative physical concepts that satisfy the independence principle. The method is simple, but powerful.

1. Decompose the design problem to determine the main FR's and select the critical FR's.
2. Generate a list of subsolutions (solution fragments) for each critical FR.
3. Visualize alternative physical concepts that maintain independence of FR's by combining different subsolutions to each FR.

Illustrative Example 1.5: Develop alternative physical concepts for a "better hand powered vegetable peeler" using the logical building block method.

Step 1: Decompose design problem and select critical FRs.

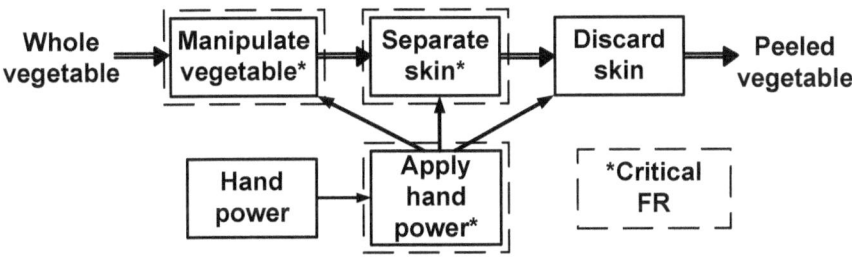

Step 2: Generate subsolutions for each critical FR.

FR1: Manipulate Vegetable	FR2: Separate Skin	FR3: Apply Hand Power
Hold in hand	Knife-blade	Reciprocate
Mount in stand	Wire	Rotate crank
Rotate on mandrel	Abrade	Rock back & forth
Tumble in bowl	Tool	Twist

Step 3: Combine subsolutions into different physical concepts.

Subsolution Combination	Alternative Physical Concept
Hold in hand + knife blade + reciprocate	Conventional peeler
Tumble in bowl + abrade + rotate crank	Sand paper lined rotating bowl
Rotate on mandrel + tool + rotate crank	Lathe like device.

Minimize Information Content

Experience has shown that reducing design complexity is a guaranteed path to good design. Nam Suh [2] has formalized this self-evident truth as the *information principle*:

Minimize information content of the design

Information content is a measure of complexity. Minimizing information content therefore maximizes simplicity. Information content is best minimized using the following common sense strategies.

1. *Eliminate* sources of information content.
2. *Simplify* by reducing the information content of the sources that remain.
3. *Standardize where possible* to further limit sources of information content and the amount of information contained in each source.

Common sense design goals that minimize information content.

Source of Information Content	Common Sense Design Goals
Number of parts.	Design to minimize the part count. Design parts that remain to be simply shaped, easily made, and assembled.
Dimensions and Tolerances	Design to reduce the number of dimensions per part and per assembly. Avoid critical dimensions that depend on the assembly of multiple parts. Relax tolerances wherever possible.
Unique features, characteristics, functional surfaces, etc. contained in a component.	Design to reduce the number of each attribute. Standardize and rationalize[4] alternatives. Develop a "design with standard features" approach.
Number of different tools and processes used in an assembly or in component manufacture.	Design to reduce the number of tools and processes. Standardize and rationalize all tools.
Number of separate activities; Number of steps per activity; Number of repetitions of each activity	Design to reduce the number of activities, steps, repetitions, etc. Standardize and rationalize.
Randomness and variability.	Specify robust parameter values. Eliminate unconstrained components. Error-proof the design.

---

[4] To "standardize and rationalize" means to (1) reduce the number of standard options used in existing designs and (2) identify the fewest number of standard options for use in <u>future</u> designs.

Illustrative Example 1.6: Which rotational component design is preferred?

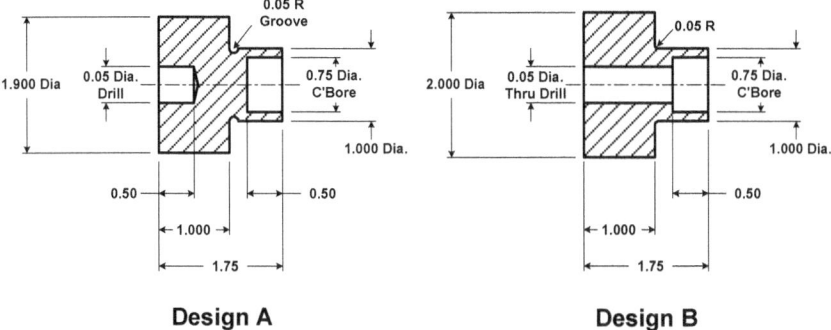

<table>
<tr><td>Design A</td><td>Design B</td></tr>
</table>

Imagine the list of instructions required to machine each alternative. The alternative requiring the fewest instructions has the least information content and is therefore the preferred design. Assuming a starting workpiece cut from 2-inch diameter bar stock and a conventional lathe, machining instructions would likely be as follows for each alternative.

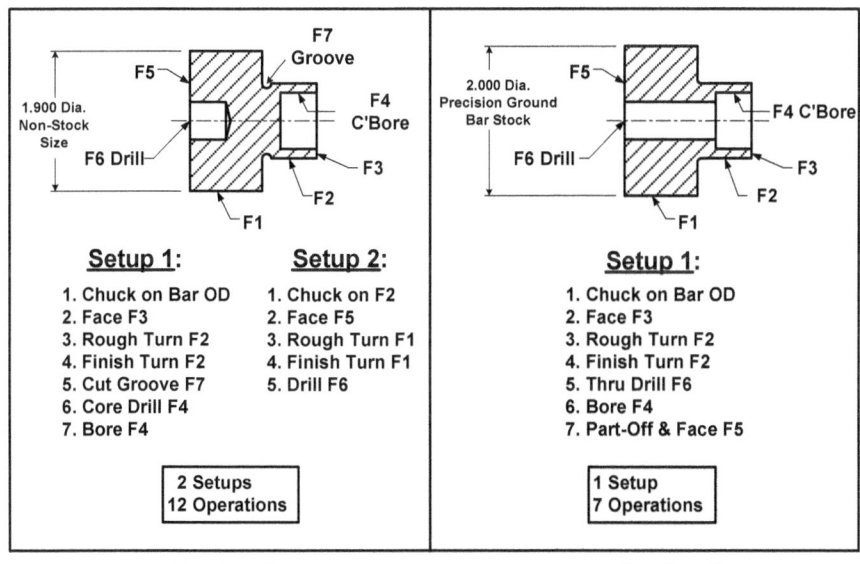

The Choice: **_Design B_** is the obvious choice because it has considerably less information content (1 setup and 7 operations compared to 2 setups and 12 operations). Note that the improvements in Design B are the result of simple common sense design choices. Specifying a standard OD allows precision ground bar stock to be used thus eliminating the need to machine the OD (surface F1). This, together with the 0.50 inch diameter through hole eliminates the need for a second setup. Finally, using a fillet radius eliminates the groove cutting operation (in Design B, the fillet radius is the cutting tool tip radius).

Illustrative Example 1.7: Skakoon [3] asks which key design is preferred, Design A, a single-sided key that is easy to make but can only be inserted one way, or Design B, a symmetric double-sided key that is easier to use (no orientation required), but more difficult to make (both sides must be cut).

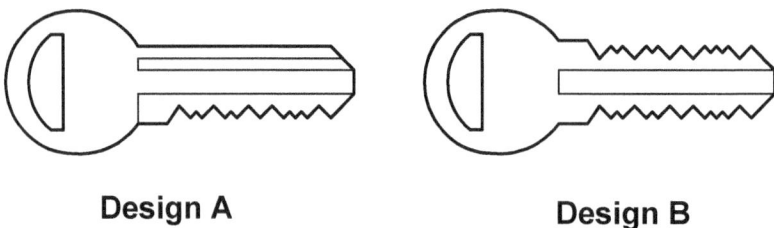

Design A   Design B

The Choice: The trade-off that minimizes over-all information content of the design while also best satisfying customer needs is preferred. An automobile key will likely be made only once, but used innumerable times making **_Design B_**, the double-sided key, the better choice. This choice requires no orientation which ensures ease of use under normal circumstances as well as in an emergency. **_Design A_**, on the other hand, may be preferred for a house key if ease of making a copy outweighs ease of use. Having an accurate and clear understanding of customer needs is always the first step.

Illustrative Example 1.8: When choosing between alternative physical concepts or part decompositions, a "complexity scorecard" that enumerates the information content of each alternative can be helpful. In addition to encouraging the team to consider all of the sources of information content associated with each alternative, such a scorecard provides insight into trade-offs and intrinsic cost.

Complexity Scorecard

Sources of Information Content	*Design A*	*Design B*
• Number of new designed parts	12	5
• Number of new vendors	3	2
• Number of unique parts	2	3
• Number of major new tools	2	4
• Number of new production processes	0	1
Total	**19**	**15**

The Choice: Since **_Design B_** has a lower "Complexity Score", i.e., less information content, it is the preferred choice.

Illustrative Example 1.9: Information content correlates with total cost. Therefore, estimating the information content of competing alternatives can be a quick and convenient way to make economic decisions as well as design decisions. To illustrate, assume 500 of the parts designated as Design B in Illustrative Example 1.6 are to be delivered to a customer by a machine shop. The step diameter (feature F2) is critical and must be machined within 1.000 ± 0.002 inches. Three machining alternatives are available: (1) engine lathe, (2) automatic screw machine, and (3) NC lathe. Which machine tool should be selected (a) based on direct cost and (b) based on information content? Unit raw material cost is $15.00 and unit salvage value is $1.00. The scrap rate, unit processing cost, and total unit cost are tabulated for each machining alternative in the table below. Unit cost is computed as follows,

$$Unit\ Cost = \frac{Total\ Direct\ Cost}{Units\ Shipped} = \frac{Raw\ Material + Processing - Scrap\ Rate \times Salvage}{1 - Scrap\ Rate}$$

Machine Tool	Scrap Rate (%)	Processing Cost Per Unit ($)	Unit Cost ($/unit)	Scrap (units)
Engine Lathe	17.9	10.00	30.23	109
Automatic	1.6	15.00	30.46	8
NC Lathe	0.0	18.00	33.00	0

(a) Choice Based on Direct Cost: Although the **engine lathe** produces a large amount of scrap, its low processing cost makes it the best choice based on direct cost. If the shop manager is concerned with scrap parts, he may also consider selecting the **screw machine** because the incremental cost penalty per part is small ($0.23/part) and the amount of scrap is much less. However, both alternatives produce some scrap so 100% inspection is required to ensure that no defective product reaches the customer. Although the NC lathe produces zero scrap, it would not be considered because direct cost is unacceptably high.

(b) Choice Based on Information Content: In addition to direct work content, estimation of information content would also include the information content involved in dealing with scrap parts (100 % inspection, material handling and storage, scrap dealer, etc.) and accidental delivery of a defective part (contract penalties, damaged reputation, etc.). The **_NC Lathe_** is the obvious choice based on information content because it produces zero scrap and therefore avoids all information content associated with scrap. In essence, using the information principle to make the choice minimizes "total cost" rather than "direct cost".

1.7 GUIDED DESIGN METHOD

The *guided design method* uses the independence principle and the information principle to guide the design in a three-step procedure:

1. Determine the critical functional requirements (FR's).
2. Synthesize a physical concept that maintains independence of FR's.
3. Simplify the physical concept by minimizing information content.

The following example, which is adapted from [2], illustrates the method.

Illustrative Example 1.10: A traditional water faucet uses two valves, one for hot water and the other for cold. Although this design works and has been used for many years, it lacks clarity because both valves must be adjusted in a trial and error process to control flow rate and temperature. Use the guided design method to develop an improved design.

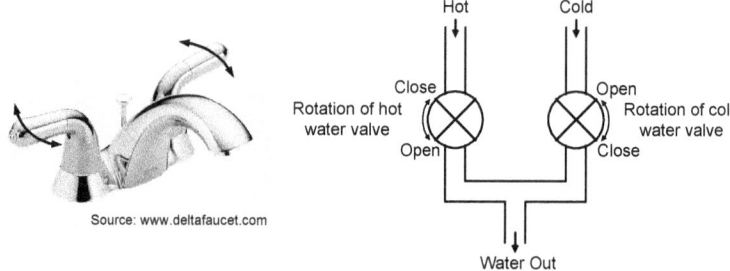

1. **Critical functional requirements:**
   FR1: Control water flow rate without affecting water temperature.
   FR2: Control water temperature without affecting water flow rate.

2. **Physical concept that maintains independence of FR's:** Control A controls water flow by opening or closing an outlet valve. Control B controls water temperature by simultaneously opening/closing a hot water valve and closing/opening a cold water valve. Control A and Control B are independent and therefore FR's are independent.

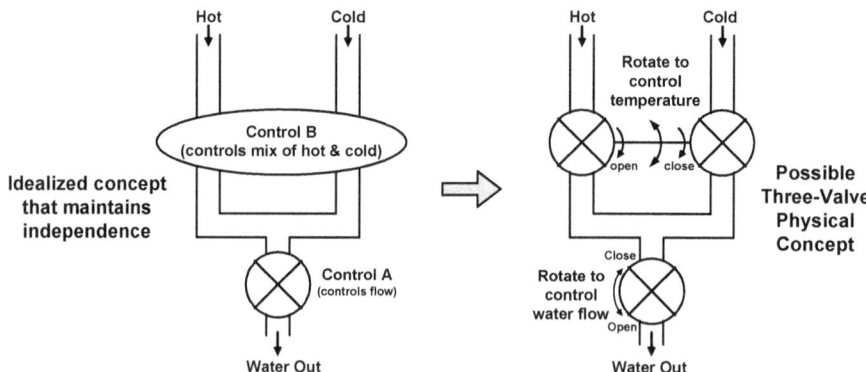

3. **Simplify by minimizing information content:** The initial physical concept is complicated and is not practical because it uses three valves while the traditional design only uses two. Applying the information principle, the initial design is simplified by first eliminating one of the valves, then replacing the two valves with two sliding plates, and finally replacing the two sliding plates with one multi-motion (sliding and rotating) plate.

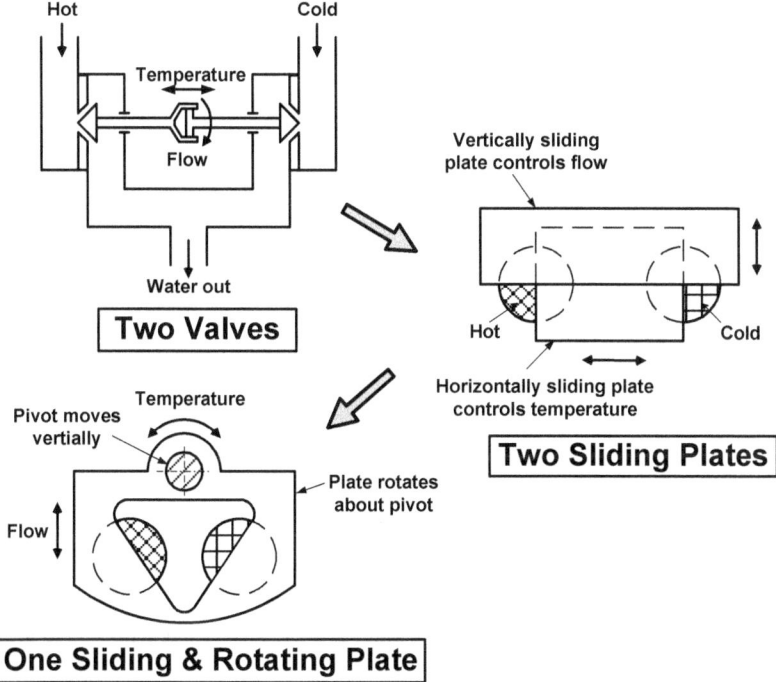

The commercially available water faucet sold by Delta (www.deltafaucet.com) shown below reduces information content even further by replacing the sliding and rotating plate with an easier to manufacture ball and socket configuration.

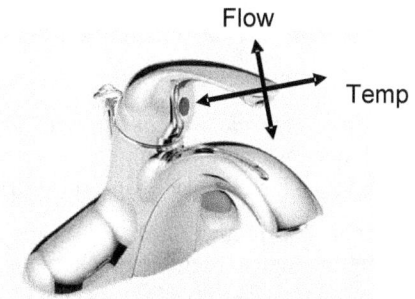

Source: www.deltafaucet.com

1.8 ADVICE, INSIGHTS, AND CAVEATS

Thinking principles. Thinking principles are ways of mentally picturing good design. They open windows in the designer's mind that allow the problem of design to be seen in a new or different light. In the Design for Everything approach, the concept of "minimizing friction" is a mental compass that points the way to good design.

Don't underestimate the importance of design. Design is a relatively minor part of the corporate budget, yet it drives most direct and indirect costs.

Do it right the first time. Fixing the design after it has gone into production can triple or quadruple cost and often, the problems can never be fully corrected because of irreversible choices that have been made. To minimize total cost and avoid sub-optimal design, spend the time and money to do it right the first time.

The earlier the better. Employ the Design for Everything approach from the beginning. One good design decision made *early* in the design equals years of cost reducing the design once it is in production.

In general, less is best. Therefore, do not be guided by piece part cost alone. The goal should be the *least* number of *simply* shaped parts, produced and assembled using the *least* number of *easily controlled* processing steps.

Standardize very carefully. Standardization is a powerful, far-reaching way to reduce information content. It may also impose unintended compromise that restricts design and marketing options in undesirable ways. See Section 7.1.

Avoid surprises; never make assumptions about customer needs. All assumptions and "judgment calls" should be customer tested and verified.

Avoid surprises, never make assumptions about manufacturability. Always obtain input and advice from knowledgeable manufacturing experts. Every design decision, if not carefully considered, can cost extra manufacturing effort and productivity loss.

Consider the price for not doing good design.
- Lost sales, returns, warrantee cost, sullied reputation.
- Time spent solving problems that should never have occurred.
- State-of-the-art technologies and methods fall short of promise.
- Extra cost and effort continues for the life of the design.

Undesirable interactions and information content are everywhere. Walk almost any assembly line, attend meetings dealing with cost over-runs and schedule slips, listen to the evening news (e.g., Toyota recall, Gulf oil spill).

Creativity can be stimulated. The number of new ideas and patentable concepts that can be generated using the logical building block method is mind boggling. A 3 x 5 matrix might yield more than 243 different concepts.

SECTION 2

VOICE OF THE CUSTOMER

All things man-made are designed. Some designs are great, others mediocre, and still others barely acceptable. This range in success can frequently be attributed to how well the design satisfies customer needs. Good design begins with "the voice of the customer". Understanding customer needs ensures that the right problem is being solved, that the design solution will be superior to competitor designs, and that there will be no surprises when the design is ultimately brought to market. In this section, the Design for Everything approach is applied to the challenge of understanding customer needs and of creating designs that satisfy customer needs in ways that delight and sustain over the long-term.

2.1 DEFINITIONS AND BASIC CONCEPTS

Client: Designs are created by clients to benefit customers. The *client* is typically a manufacturing enterprise seeking to sell products and equipment, a developer or business seeking to provide a service, or a government entity charged with procuring military hardware and providing public infrastructure. The client profits from good design because of enhanced reputation, increased demand, and higher selling price combined with lower manufacturing and life-cycle costs.

Customers: Any entity that interacts in any way with the design is a customer. This includes manufacturing, distribution, use, and disposal as well a regulation, testing, and code authorities. Customers benefit from good design because it meets their needs in a reliable, easy to use, enjoyable, and safe way.

User: The end customer for whom the design is intended. User needs are primary customer needs and usually have the highest importance weighting. Typically the end user benefits most from the design's functionality, appearance, ease of use, reliability, long life, and safety, but other types of customers such as installers and service technicians also benefit. To be acceptable to end users, the design must meet well defined and clearly understood needs in ways that delight and ensure sustainable satisfaction. For other customers such as distributors, middle-men, and so forth, the design should enhance their business, reduce their cost, and contribute to their overall success.

Need: Any quality, attribute, or trait of a potential design that is desired by customers. Needs are synonymous with *voice of the customer* and include both "must haves" and "wants".

Voice of the Customer: Customer needs as expressed by actual customers, not imagined by the design team, marketing, or other client entities.

Hear the Voice of the Customer: To hear the voice of the customer, it is necessary that the design team develop a comprehensive understanding of needs through direct interaction with real customers. This requires that the team conduct individual customer interviews, observe customers in action in the use environment, and when appropriate, gain first hand experience with the use environment themselves.

Design Specifications: Needs are expressed in the "language of the customer" as opposed to design specifications, which define in precise, measurable terms *what* the design is to do expressed in the "language of the engineer". Specifications do not tell the team how to address customer needs, but they do represent an unambiguous agreement on what the design must achieve to satisfy customer needs [4]. Think of the design specification as a formal statement of functional requirements stated in measurable terms and accompanied by specified limits and target values.

2.2 PROBLEM STATEMENT

Mission Statement: Design projects typically begin with identification of a particular market opportunity or un-met user need together with broad constraints and objectives for the project. This information is frequently formalized as a *mission statement* (also sometimes called a *charter* or a *design brief*). The mission statement is typically the result of product planning activities within the firm and it usually specifies the direction to move in but not the destination. Ulrich and Eppinger [4] suggest that the mission statement should include a brief (one-sentence) description of the design, key business goals, target market(s) for the design, and assumptions and constraint that guide the design effort.

Problem Statement: In the Design for Everything approach, a first step in understanding customer needs is to clearly formulate a problem statement for the design effort. Building on the mission statement if one is available, the *problem statement* is a clear and concise declaration of what the design is to achieve together with carefully chosen focusing assumptions. *Focusing assumptions* limit the scope of the problem. For example, including the term "manually powered" eliminates other means of powering the design from consideration. Such assumptions constrain the design solution and must be carefully chosen to be in harmony with the mission statement and other stated goals of the project.

Purpose: The problem statement establishes the level of the problem and forces the team to carefully consider exactly what the design task involves. It tells the team what to focus on and what it is about the use environment that needs to be understood. Most importantly, the problem statement provides a context for approaching customers by suggesting a list of questions that need answering.

Open to Change: The problem statement should constantly be questioned to insure that the right problem is being solved. As the team interacts with customers and searches for solutions, the nature of the problem may change. For example, the problem may morph from "design a new, innovative cooler" to "modify the current cooler design so that it is easy to transport". Being sensitively tuned to the true underlying needs of the design problem is one of the closely held secrets of good design.

Solution Independent: It is very important that the problem statement not suggest or imply a design solution or otherwise misguide the problem solving process in any way. *How the problem is defined determines the solutions that are proposed.* If the problem statement suggests a solution, then the opportunity for finding the best solution as well as for innovation is lost.

2.3 CUSTOMER TYPES

Who purchases the design? Who uses the design? Who services and maintains the design? Who manufactures the parts that go into the design and who assembles the parts to form the finished design? Who else comes in contact with the design? Who regulates the design? When a design is viewed in terms of these questions, we see that there is often many different types of customers for the design whose needs must be considered when deciding what the design must do and how it is to do it.

Production-Consumption Cycle: Each stage is a customer.

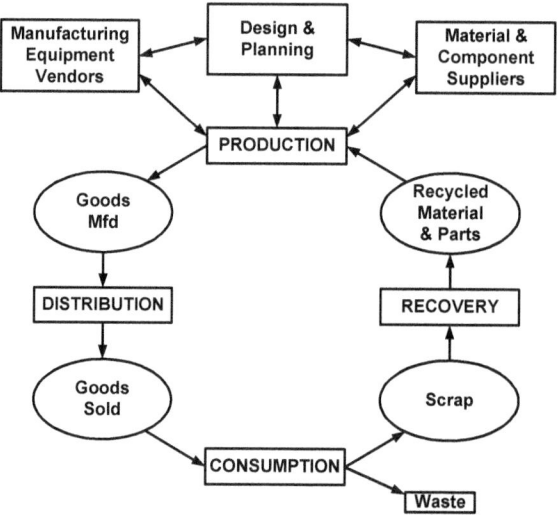

Distribution Channel: Each organization or individual in the distribution channel is a customer.

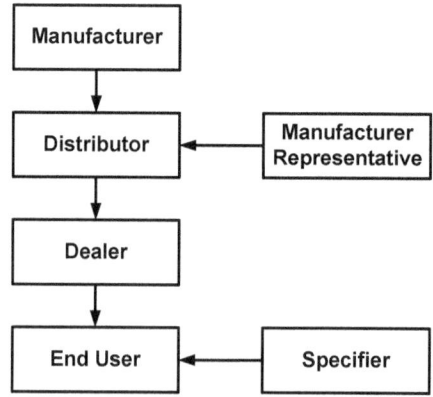

Order-Flow Process: Each organization or individual involved in the order-flow process is a customer. A typical order-flow for an electrical distribution product such as a circuit breaker panel is shown below.

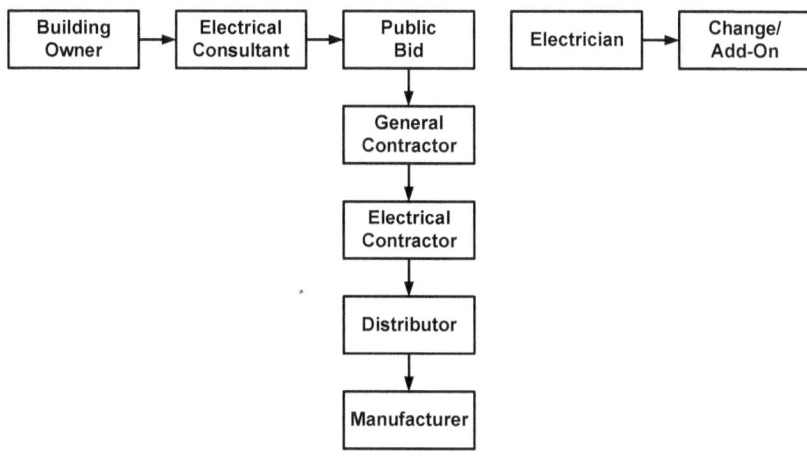

2.4 GATHERING CUSTOMER NEEDS[5]

Gather: Customer needs for each type of customer are determined by interviewing individual customers, observing users in the use environment, conducting and observing focus groups, sending out surveys, observing similar or competitive products in use, and so forth.

Interpret Data and Organize: Translate the data into concise statements of need. Group the need statements according to underlying similarities to define a manageable list of primary customer needs that can be used to guide the design process. To the extent possible, develop a sense of relative importance of each primary need.

Verify: Obtain customer feedback on the validity of primary customer needs, relative importance's, and any assumptions that might have been made. Anything short of asking the customer could ultimately result in an unpleasant surprise.

Refine: Gather and interpret additional data as needed to complete an accurate picture of customer needs.

---

[5] See reference [4] for more in-depth coverage of this important topic.

2.5 HARMONIZE CN'S AND FR'S

The diagram below shows the relationship between customer needs (CN's) and functional requirements (FR's). This diagram is often called a *house of quality* because it resembles a house with many rooms. Each room contains valuable information that is developed by creating the diagram.

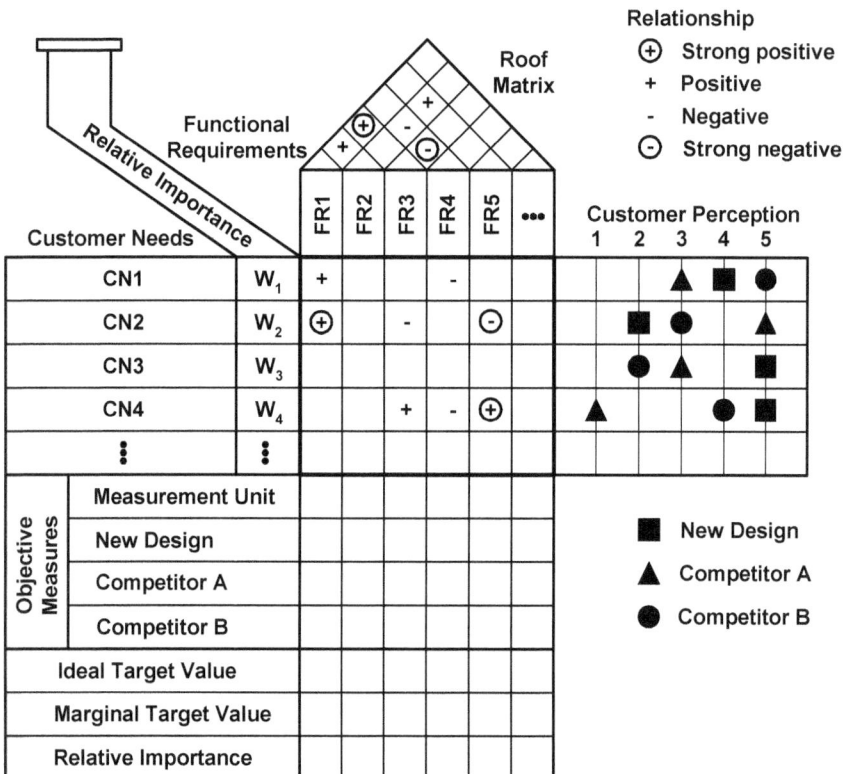

Relationship Matrix: The *relationship matrix* shows the relationship that exists between customer needs and the functional requirements. If a functional requirement affects or relates to a customer need, the strength (strong or mild) and nature (positive or negative) of the relationship is shown at the intersection of the row and column. In general, each customer need should be associated with one or more functional requirements. It is possible that a particular functional requirement, such as **FR2** in the diagram above, does not relate to any customer need. This is an indication that the functional requirement may not be needed. Customer need **CN3**, on the other hand, is not associated with any functional requirement. This tells the team that an additional functional requirement is desirable to ensure that this need is satisfied by the design.

Roof Matrix: Shows positive and negative relationships that may exit between the various functional requirements. For example, we see that an improvement to **FR3** will strongly benefit **FR1** whereas an improvement to **FR2** will benefit **FR1** to a lesser extent. We also see that an improvement to **FR3** will negatively impact **FR4**, but not too severely because the relationship is not judged to be strong. The roof matrix, together with an objective assessment of the relative importance of the customer needs and functional requirements, helps guide the team in making design tradeoffs and in deciding which features to emphasize and which to sacrifice.

Customer Perception: The house of quality also provides a means for comparing proposed new designs with competitor designs and other benchmark designs. The *customer perception* comparison graph (right side of the house) shows how the proposed new design compares with existing designs. Each design is rated subjectively on a scale of 1 to 5, with 5 being best (note that any convenient scale would suffice). From the diagram, it is seen that the new design is somewhat deficient with respect to customer need **CN2**. The team may choose to improve the new design to better satisfy this need or to ignore the deficiency depending on the relative importance of **CN2**. In theory, market success is ensured when the new design is perceived to be better than best-in-class competitor designs with respect to all of the customer needs.

Objective Measures: The *objective measures* portion of the house (i.e., basement) allows the team to benchmark the proposed new design against competitor designs and other benchmark designs with respect to numerical measures of the functional requirements. This evaluation provides insight into how the designs compare technically. It also shows how numerical values of the functional requirements correlate with customer perception of the different designs. This can be very helpful in setting the marginal and ideal target values for the functional requirements. The "marginal value" determines acceptability and is expressed as the minimum or maximum acceptable value for the functional requirement. When possible, marginal values should be precisely specified because they determine acceptability. The "ideal value", on the other hand, is the value that would be most desirable or the best that can be hoped for. Ideal values can be specified precisely or in fuzzy terms such as "about" or "as low as possible" or "as high as possible". They can also be specified in ranges.

Procedure: There are no hard and fast rules for constructing and using the house of quality. It is important to remember that the goal is to clearly understand and define what the design is to achieve. The "house of quality" helps by explicitly delineating the relationship between the "voice of the customer" (i.e., customer needs) and the "voice of the engineer" (i.e., functional requirements). It also provides a convenient and simple way to compare competing physical concepts.

2.6 USER-CENTERED DESIGN

User-centered-design seeks to identify and evolve the best physical concept for a design through direct interaction with users and the use environment. A multi-step iterative process for understanding user needs, visualizing physical concepts to satisfy those needs, and obtaining user feedback to refine and validate design decisions made along the way is used.

1. Formulate problem statement
2. **Understand** end user needs and use environment.
3. **Visualize** possible design concepts for solving the problem.
4. **Experiment** by testing promising design concepts with end users.
5. **Refine** based on feedback, combine best features into one or more winning concepts.
6. Repeat process as necessary.

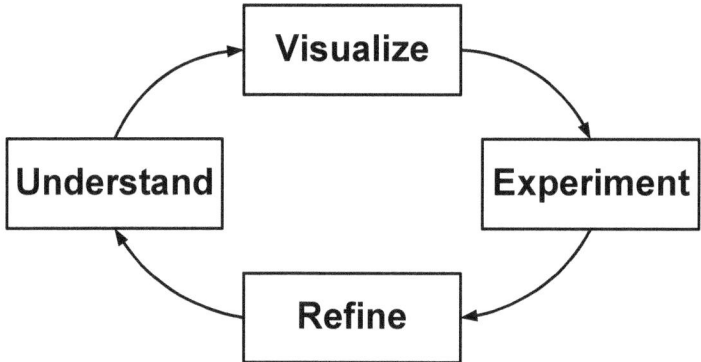

Understand: Using the problem statement as a guide, understand customer needs by interviewing users and observing the use environment. Meet with users to identify and appreciate problems from the user perspective. In addition, carefully observe the use environment, either passively or actively, to discover details that even users are unaware of. Observation can be done passively, such as watching users in action from a distance, or it can be done actively by working side-by-side with a user. Active observation has the great advantage of allowing team members to gain hands-on experience with the use environment and with using existing design solutions. Observations should be carefully documented by taking notes, making audio and/or video recordings, and taking still photographs. Photographs and videos teach a lot about user needs. For example, by displaying a large number of photographs taken of people transporting food coolers at beaches, picnic grounds, tail-gating parties, etc., the team saw the design solution (mount wheels on the cooler) emerge before their eyes as they analyzed their collective field data.

Visualize: Imagine possible design solutions. Use idea stimulating techniques such as brainstorming and logical building blocks (Section 1, page 13). Reverse engineer existing solutions and imagine how they could be improved. It is not unusual for the simple process of experiencing the use environment, observing users in action, and participating first-hand with users to stimulate design ideas and solutions by imagination and ingenuity alone. Select among the many ideas generated by winnowing obviously impractical ideas, grouping similar ideas, and combining the best features of different groups of ideas.

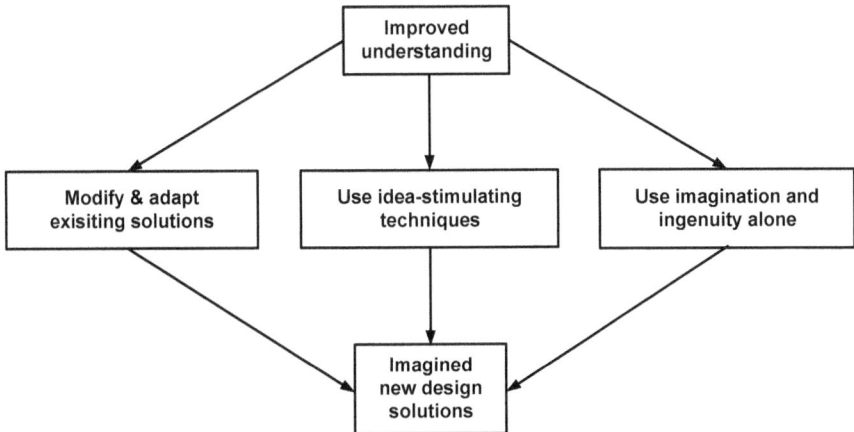

Experiment: Experiment with the various ideas by building simple models and mock-ups. Narrow the ideas to the best concepts and test with users. A good way to do this is to construct several models (in practice, 3 is a good number) and include different mixes of promising ideas and innovations in each model. Take the models to the field and let users experience them. Keep track of what the user likes and dislikes about each model. By doing this, the team gets an idea of what design features and innovations are important to the user and hopefully avoids making incorrect assumptions about what they think users would like.

Refine: Based on feedback and insights gained from the experiment step, combine the features and innovations that customers like and eliminate those they don't to define a final definitive model of the physical concept. Take this back to the field and verify, again by letting users evaluate the model, that the refined physical concept is the best one possible.

Repeat Process as Necessary: It may be necessary to repeat the process two or more times. The first iteration establishes a design direction and the follow-on iterations add detail until enough is known to be sure the physical concept is a "winner" and that user-related design knowledge is complete. Always remember, the goal is to avoid unfounded assumptions and surprises.

2.7 ADVICE, INSIGHTS, AND CAVEATS

Understand the needs of all customers. This frequently provides design knowledge that can be gained in no other way.

Interview real customers. Do not rely on focus groups or surveys alone.

Stimulate discussion. Bring props (existing product, competitor product, model, prototype, etc.) to help stimulate discussion and reveal problems.

Test all ideas and concepts with real customers. The whole point of "hearing the voice of the customer" is to avoid making incorrect assumptions about customer needs and to avoid unpleasant surprises when the design reaches the marketplace. This requires early and plentiful feedback from real customers.

Observe more than ask. When interviewed, end-users tend to sometimes second guess the answer being sought. In addition, it is not unusual for customers to take for granted the "work-around" and "coping mechanisms" that they have developed or use regularly. In many cases, customers simply don't think of problems when interviewed because they are so use to them. This is especially true on the manufacturing floor, where workers perform repetitive tasks day-in and day-out. Experience has shown that "feeling the pain" through close and careful hands-on observation is the single best and most efficient way to accurately understand customer needs.

Identify lead users. Lead users experience needs ahead of others and typically stand to benefit greatly from innovation and improved performance and functionality. These users can be a valuable source of customer needs as well as excellent judges for testing design concepts and the validity of assumptions.

Look for latent needs. Latent needs are needs that are currently unfulfilled or have not been recognized. Designs that are first to satisfy latent needs are often highly successful.

Understand the need behind the how. Sometimes customers express preferences in terms of "how". Be sure to understand the "need" that underlies the "how" because it is likely that a better "how" is possible.

Capture the use environment. Video and still photography teach a lot about customer needs. Such documentation can also come in handy during concept generation and for making presentations.

Quantity equals quality. The probability of identifying the best idea increases exponentially with the number of ideas generated. Explore the whole design space by encouraging wild ideas and deferring judgment until later.

SECTION 3

FORCE AND DEFORMATION

Engineering designs must often support and/or transmit force. Force can arise because of external loading developed during use or manufacture or because of effects such as weight, inertia, temperature change, preload, etc. If not properly designed to resist the effect of force, components and structures will fail, either because the internal stress created by the force exceeds the strength of the material or because the component deforms to the point where it can no longer perform its function. This section explores how the Design for Everything approach can be used to help identify and avoid the undesirable effects of force and deformation.

3.1 FAILURE PREVENTION

Design decisions based on the rules of good design lead to inherent failure prevention. Understanding of needs shows the way to explicit specification of operating conditions and environmental factors such as anticipated loads and service life. Predictable and unambiguous behavior clearly reveals failure modes and helps ensure robustness against hard-to-control variation. Simple components and assemblies have fewer failure modes and less opportunity for error or malfunction. A consistent focus on safety creates a redundant multi-link safety chain that ensures component safety, functional safety, operator safety, and environmental safety.

Factor of Safety: *Failure* occurs when a design is unable to perform its intended function. Load carrying members can fail in different ways or *modes* that depend on material (ductile or brittle), type of loading (steady, cyclic, repeated, impact), and environmental conditions (temperature, reactive atmosphere). Common failure modes include:

1. Failure by elastic deflection
   - Excessive change in shape (sag, twist, etc.) caused by static load
   - Excessive vibration amplitude
   - Buckling
2. Failure by plastic deformation (yielding)
   - Ordinary (room) temperature
   - Elevated temperatures (creep)
3. Failure by fracture
   - Sudden fracture of brittle materials
   - Fracture of flawed or cracked members
   - Fatigue (progressive) fracture
   - Fracture with time at elevated temperature

The *factor of safety* directly protects the design against uncertainties associated with material properties, force magnitudes, operating conditions, and other hard-to-control variation. For safety, the working value P_w associated with a particular failure mode must be less than the critical value P_f. The *factor of safety* (*FS*) is defined as the ratio of critical value to working value, i.e., $FS = P_f/P_w$.

Fail-Safe Design: This design philosophy seeks to design in such a way that, if failure should occur, the failure will be economically acceptable and will not harm humans or the environment. Fuses, shear-pins, and other replaceable elements that are designed to fail are examples. When failure and/or danger cannot be prevented directly, indirect methods such as special protective systems can be used. A safety switch that automatically disconnects electric power if a heater is accidentally tipped illustrates this approach.

3.2 DESIGN EFFICIENT FORCE-FLOW

Force-flow is the path taken by a force as it passes through a component, structure, or assembly of parts. Efficient force-flow maximizes capacity and minimizes material waste. Generally, force-flow can be determined by inspection using common sense. The figure below illustrates force-flow through a C-clamp. When the screw is tightened, various modes of loading occur in different regions of the C-clamp. Following the force-flow loop, we see that there is compression in region A and bending and tensile loading in region B. In region C the screw threads are loaded in shear as force flows from the curved beam into the screw. Closing the loop, region D is loaded by compression. Assuming the C-clamp is made of mild steel, it could fail by compressive yielding in regions A and D, tensile yielding in region B, and yielding due to shear of the internal screw threads in region C.

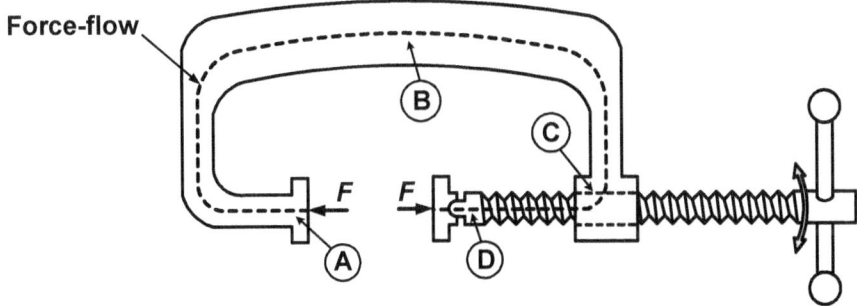

Develop a Balanced Design: The failure mode having the lowest factor of safety is critical. To balance the design, select proportions so that the factor of safety is about the same for each failure mode. For the C-clamp, region B is critical because it is subjected to both tension and maximum bending. The design is balanced when the C-clamp is proportioned such that the factors of safety at A, C, and D are all about the same as that at B.

Material in the uniform cantilever beam below is under utilized (*FS* = variable) while the tapered beam is balanced (*FS* = constant). Both beams carry the same force, but the tapered beam uses half the material.

Uniform cantilever beam	Tapered cantilever beam
$FS = \infty$ *at* $x = 0$	*for* $0 \le x \le L$
$FS = \dfrac{S_y\left(bh^2\right)}{6LF}$ *at* $x = L$	$FS = \dfrac{S_y\left(bh^2\right)}{6FL} = constant$
Volume $= bhL$	*Volume* $= \dfrac{1}{2}bhL$

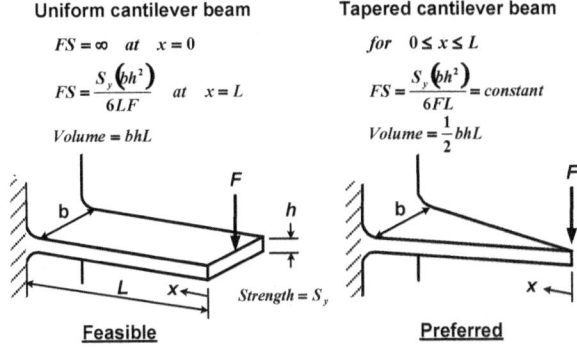

Feasible	Preferred

Use Short and Direct Force-Flow Paths: Design so that the force-flow follows the shortest most direct path possible. Always avoid changes in the direction of the force-flow when possible.

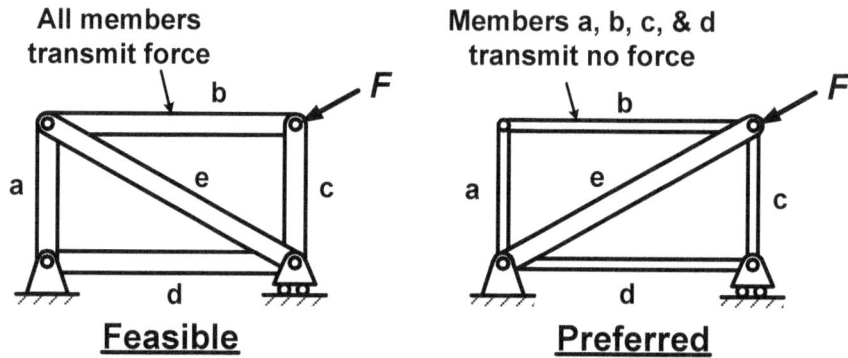

Avoid Disruption of Force-Flow: Visualize force-flow as parallel lines of flowing fluid. The notch in the tensile member shown below disrupts the force-flow. Force makes the flow-lines want to straighten, causing them to bunch and concentrate at the surface of the notch. This **weakens** the bar by introducing *stress concentration*, i.e., a region where the maximum stresses (σ_{max}) can far exceed the average stress (σ_{avg}).

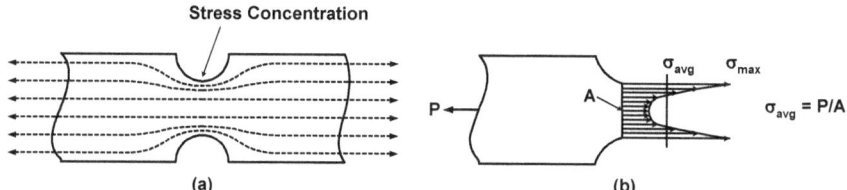

Smooth the Force-Flow: The severity of stress concentration can be reduced by visualizing various geometry modifications that smooth the force-flow and avoid bunching of the flow-lines. As illustrated below, the designer could (a) remove material by widening the notch; (b) if the notch profile must be preserved, addition of adjacent smaller notches would help; or (c) if the surface contour surrounding the notch must be retained, then drilled holes could be considered.

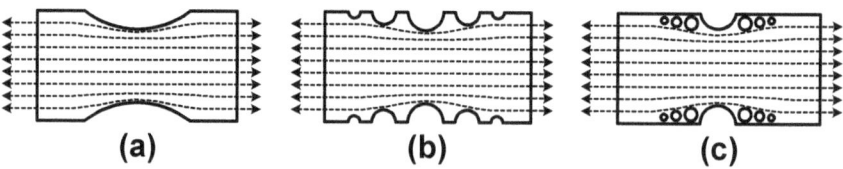

Design Gradual Transitions: When change in the direction of the force-flow cannot be avoided, design so the force-flow changes direction gradually and smoothly.

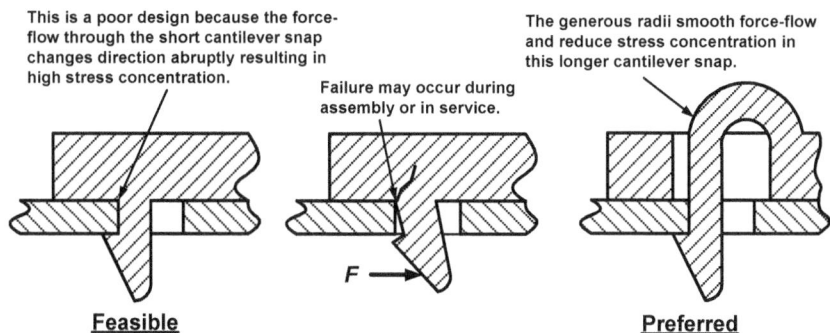

Avoid Stiffness Mismatch: When force-flow transfers from one component to another in an assembly, severe stress concentration can occur if there is a mismatch in the stiffness of the two parts. For example, the stress concentration factor for a parallel fillet weld is approximately twice that of a transverse fillet weld because of stiffness mismatch. Stress concentration can be avoided by designing in such a way, that under load, adjacent parts deform in the same sense and, if possible, by the same amount [1]. This is illustrated by the shaft and support assembly shown below. By making the support relatively flexible in the region where force-flow transfers, the force-flow in the preferred design is smoother and more gradual, resulting in lower stress concentration.

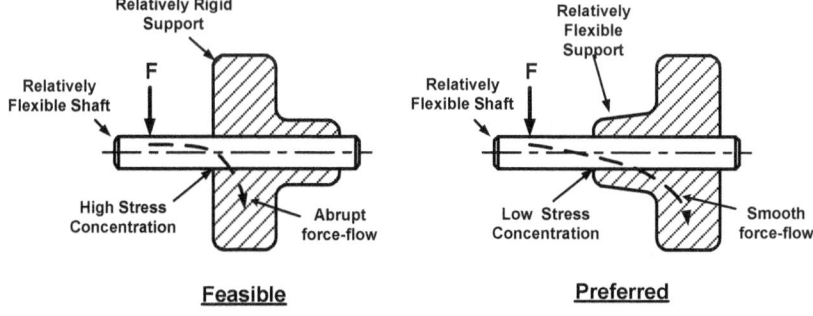

3.3 AVOID UNDESIRABLE DEFORMATION

Real materials deform when subjected to force. This deformation can interfere with function and cause undesirable interactions that degrade capacity and ease of manufacture. When performance and manufacturing difficultiess are hard to explain and force-flow is present, suspect unwanted deformation. Visualization of the force-flow and consequent deformation will usually reveal the problem.

Illustrative Example 3.1: The gear pump shown below works by trapping fluid in the spaces "S" between the teeth of meshed spur gears and moving the fluid from the intake side to the discharge side. The spaces "S" are formed by the gear teeth, the housing, and the end-plates so fluid cannot escape. Since only a small fraction of the fluid returns through the gear tooth mesh at "M", high pressure develops when rotation of the gears forces the trapped fluid to flow against a discharge resistance. The design shown has failed to achieve specified discharge pressure due to fluid leakage between the end-plates and the gear faces. "Beefing up" the end-plate thickness has not helped.

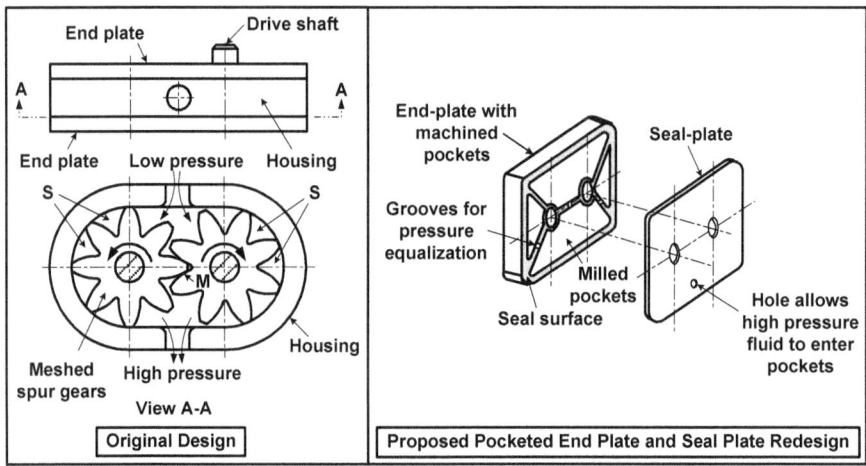

The Problem: An undesirable interaction exists between the "seal pressure" and "hold pressure" functions, both of which are fulfilled by the end-plates. High fluid pressure deforms the end-plates which allows high pressure fluid to flow back between the gear faces and end-plates to the low-pressure intake side of the pump.

The Redesign Fix: The end-plate is used to "hold pressure" and a separate seal-plate, inserted between the end-plate and the housing, is used to "seal pressure". Initial sealing is provided by the seal-plate pressing as a spring against the gear face. A hole drilled in the seal-plate allows high pressure fluid to flow via interconnecting grooves into pockets that are milled in the end-plate. The high fluid pressure helps press the seal-plate against the gear face. In essence, the seal-plates eliminate the leakage path, decouple unavoidable deformation of the end plates from pump performance, and provide a sealing effect that increases in proportion to output fluid pressure.

Visualize Force-Deformation Interactions: Follow the force-flow path and imagine how the component, structure, or assembly will deform under the action of the forces being applied and transmitted. Visualize the impact of this deformation on the functionality of the design and design to eliminate or mitigate undesirable effects. As illustrated below, when possible, design to have the force-flow supplement functionality rather than adversely affect it.

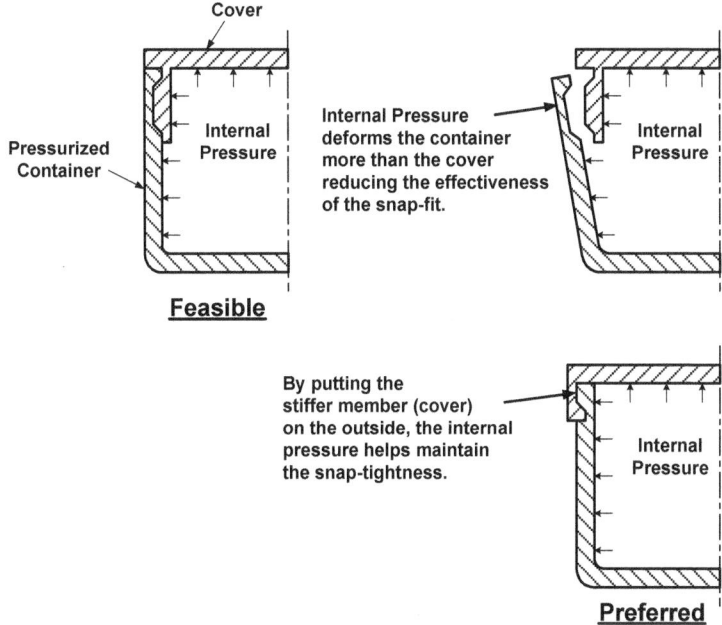

Use Exact Constraint: Exact constraint reduces undesirable and hard to control deformation due to assembly and service. An object is exactly constrained when just enough constraints are applied to unambiguously define its position. Consider the 3-2-1 fixturing principle. Three orthogonal planes, A, B, and C are used to locate the object by having the object first contact three points on plane A, then two points on plane B, and finally one point on plane C thereby exactly eliminating all six degrees of freedom (three translations and three rotations). Components that are located this way are easy to assemble, they don't bind or move unpredictably, and their position is predictable and repeatable. See reference [3] for further elaboration of exact constraint design.

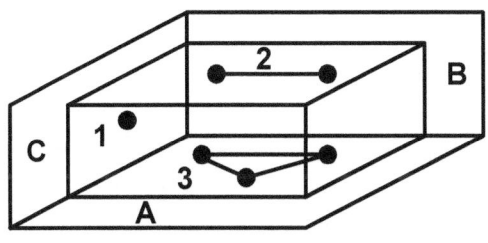

Design to Make Over-Constraint Predictable: In many assemblies and structures such as weldments and spot-welded structures, redundant force-flow paths may exist. A *redundant force-flow path* is one that could be removed and still leave the assembly or structure in static equilibrium. For redundant force-flow, *force divides in proportion to the stiffness of the load path.* Therefore, redundant force-flow paths make deformation of components and assemblies unpredictable and ambiguous unless stiffness of each force-flow path is controlled.

When rigidity and stability needs require over-constraint, do the following:

1. **Design-in predictable stiffness and strength.** Design so that stiffness (elasticity) and strength of each redundant force-flow path is known. When possible, make the strength of each path approximately proportional to its stiffness.

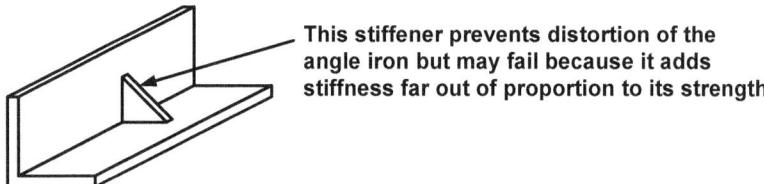

This stiffener prevents distortion of the angle iron but may fail because it adds stiffness far out of proportion to its strength

2. **Design so that one force-flow path is more rigid than the others.** Prevent residual stress in built-up structures by making one force-flow path relatively stiff compared to all others. This will help prevent unwanted distortion because shrinkage and other dimensional changes are absorbed by the less stiff, more flexible force-flow paths.

3. **Design the most rigid member to also be the most dimensionally precise.** When there are redundant load carrying members, the more flexible members deform to accommodate the smaller deformation of the most rigid member. Therefore, in "skinned" structures such as automobile bodies and refrigerators, the skin conforms to the dimensions of the stiffer, underlying frame. Make the underlying more rigid frame accurate and the assembly will be accurate.

Illustrative Example 3.2: Cross-sectional views of three alternative part decompositions for a document transport mechanism are shown below.

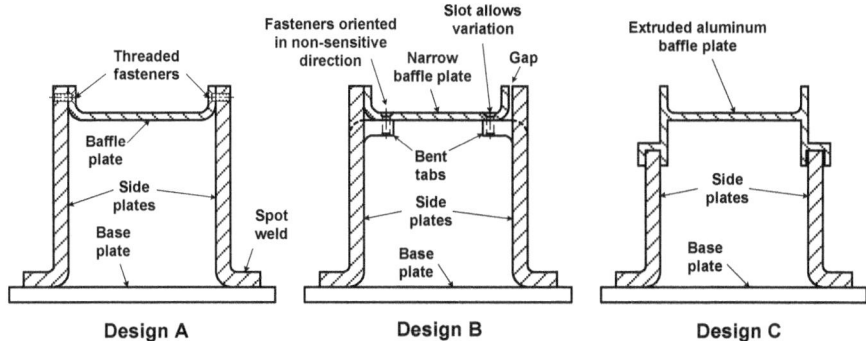

In Design A, over-constraint causes an undesirable interaction between the baffle plate and the side plates. When the width of the baffle plate is too large, the side plates are spread apart. When the width is too narrow, the side plates are pulled together by the fasteners. In both cases, deformation of the side plates interfere with proper operation of the mechanism by causing binding of the drive roller shaft (not shown), which is mounted in bushings located in the side plates. Design A should be avoided because the width of the bent metal baffle plate and precise position and angularity of the spot welded side plates are fundamentally hard-to-control and are likely to be problems regardless of dimensional tolerances or the amount of care taken during manufacture.

The Choice: Both Designs B and C avoid the undesirable interaction, but with different distributions of information content. In Design B, the baffle plate is mounted using exact constraint and is sized to always be narrower than the spacing between the side plates. In addition, the threaded fasteners are oriented so that they do not undesirably distort the side plates when tightened. In Design C, the multi-part baffle plate assembly is integrated into a single extruded aluminum part. Design B is simpler to implement and avoids costly tooling. Design C reduces part count and isolates critical dimensions into a single part (the roller shaft bushings are located in the baffle plate), but tooling is more expensive. If time and money are short or production quantity is low, **<u>Design B</u>** is preferred. If volume is high, **<u>Design C</u>** is the better choice. The most prudent choice might be to employ the principle of continuous improvement by using Design B initially and implementing Design C when demand justifies the tooling investment.

Close the force-flow path: Locally-closed force-flow paths limit and control deformation. The need for heavier construction or for reinforced bearing and transfer elements is also often avoided. Pahl and Beitz [1] recommend two proven approaches: balancing elements and symmetrical layout. As illustrated by the alternative cone clutch designs shown below, balancing elements work best for relatively small and medium forces while a symmetrical layout is preferred for large forces.

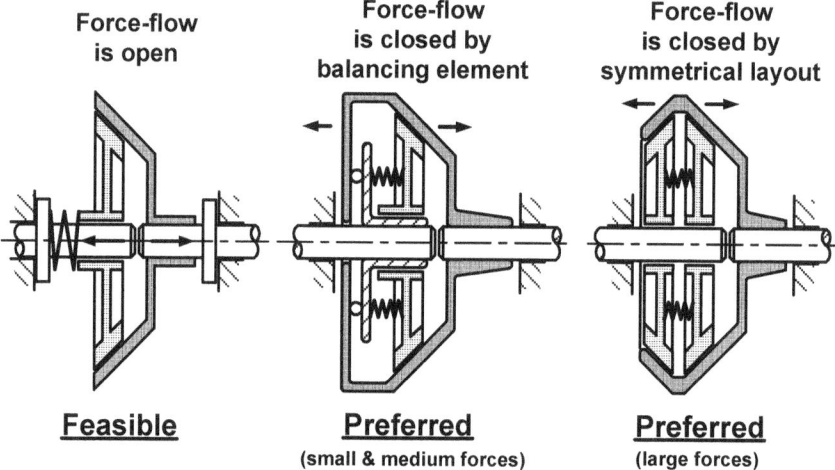

The partially disassembled mechanism shown below illustrates the use of a sub-frame to locally close the force-flow path of a preloaded spring. Prior to the sub-frame, assembly was difficult because the spring, which is used to tension the armature mechanism, caused the assembly to behave like an unstable "mouse trap" that could explode at any time during assembly.

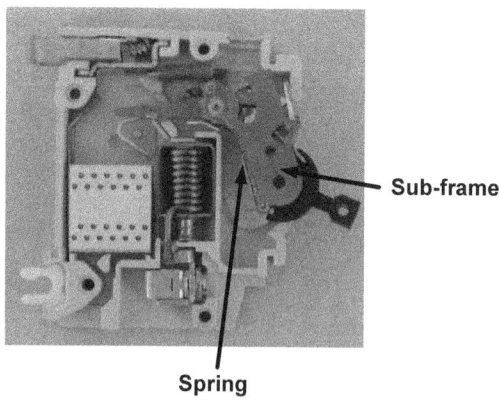

3.4 ADVICE, INSIGHTS, AND CAVEATS

Deformation causes undesirable interactions. In engineering design, it is probably safe to say that 90% or more of unexplained performance and production problems are caused by unwanted deformation. Red flags include the need to tighten tolerances, blame suppliers, or work weekends until the problem is solved. Visualize the force-flow and consequent deformation and you will discover the cause of the problem.

Favor compression loading. When there is a choice, compression loading is preferred compared to tension. Many materials are stronger in compression. Compression members typically resist fatigue failure because cracks close and don't grow under compressive stress. Force application is simple and redundant force-flows as well as stress concentration are more easily avoided. Stability is the major advantage of tension members. A member that is slightly curved will straighten under tensile load, but might buckle under compressive loading. ***WARNING***: when using compression, always guard against buckling failure.

"The ideal product has a part count of one"

Bart Huthwaite
Huthwaite Workshops

SECTION 4

PART DECOMPOSITION OPTIMIZATION

Part decomposition optimization seeks to achieve the next plateau of cost reduction by reducing manufacturing and assembly complexity and quality risk. In this section, we focus on optimizing the part decomposition with respect to "ease of manufacture and assembly". The goal is to eliminate as many parts as practical and then design the parts that remain to be easy to manufacture and assemble. The Design for Everything approach is applied by minimizing information content using the "eliminate, simplify, and standardize where possible" strategies to both guide and test for good design.

4.1 OPTIMIZATION GUIDELINES

The part decomposition is optimized for manufacture and assembly when information content is minimized using the "eliminate, simplify, and standardize where possible" strategies. To optimize the part decomposition, separate parts, processes, operations, operation steps, motions, and work directions, together with ambiguity, randomness, and uncertainty are eliminated. Parts that remain are simplified by providing design features and characteristics that make manufacture, assembly, service, maintenance, and recycling easy to do. Where possible, design options are narrowed to a subset of standard options. Over time, the number of standard options is further rationalized to reduce the options used in future designs. The following guidelines offer a disciplined and systematic guided common sense approach for optimizing the part decomposition.

1. Design to reduce part count and part types.

2. Avoid separate fasteners.

3. Design parts for easy fabrication.

4. Design for easy assembly.

5. Design parts for easy insertion.

6. Design parts to be self-assembling.

7. Design parts for easy handling.

8. Error-proof the design.

9. Eliminate randomness.

4.2 REDUCE PART COUNT AND PART TYPES

Always seek the minimum number of simply shaped components. The ideal design has a part count of one. Fewer parts mean less of everything. A part that is eliminated costs nothing to design, change, make, assemble, move, handle, purchase, store, rework, repair, or replace. Eliminate parts by (1) identifying physical concepts that require few parts, (2) consolidating separate parts into integral designs, and (3) standardizing part designs.

Theoretical Part: A part is a *theoretical part* if it receives a "yes" answer to at least <u>one</u> of the following <u>critical</u> questions:

1. Does the part move relative to other parts?
2. Must the part, for good reasons, be made of a different material?
3. Does the part need to be separate for reasons of assembly or service?

Theoretical Minimum Part Count: The total number of parts in a design that receive a "Yes" to at least <u>one</u> of the three critical questions.

Candidate for Elimination (CFE): A part that receives an answer of "No" to <u>all</u> three of the critical questions is a CFE. In other words, a CFE is a part that does not need to be separate for reasons of motion, material, assembly or service.

Efficient Part Count Range: The "ideal" part count optimally balances conflicting requirements relating to manufacturing cost, manufactured quality, development cost, and development time. The ideal may be larger than the theoretical minimum part count if development cost and/or development time are dominant project constraints or if achieving the minimum part count is not practical technically.

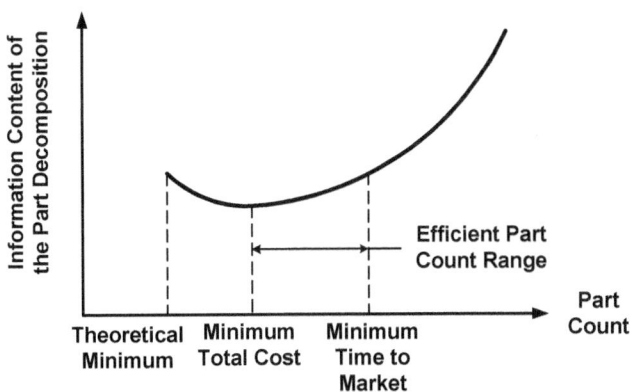

Indications that part count reduction has gone to far include:
- One or more parts are excessively heavy and hard to assemble.
- Tooling cost and/or lead times for one or more parts is excessive.
- Processes are hard to control or exceed best practice limits.
- Simpler options are rejected for hard to justify reasons.

Identify Physical Concepts that Require Few Parts: Use the "guided design method" (page 18) to identify physical concepts that satisfy customer needs in the best way possible, maintain the independence of functional requirements, and minimize information content. If more than one acceptable physical concept is identified, then evaluate each concept with respect to its theoretical minimum part count and the complexity of each part. The alternative having the *least number of simply shaped theoretical parts* should be chosen. This is the most effective way to optimize the part decomposition and ultimately, to minimize intrinsic cost of the design.

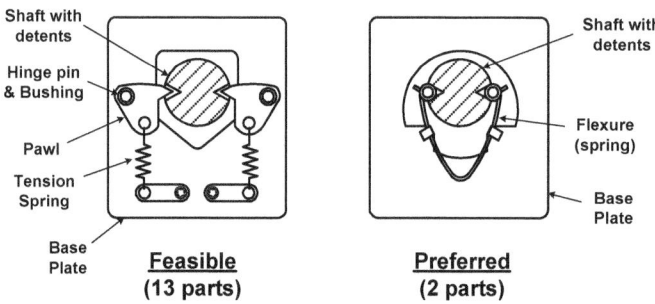

Develop an Integral Design: Integral design is the consolidation of several candidates for elimination (CFE's) into a single part. Do the following:

- Examine CFE's jointly and individually. Develop alternative integral designs. Remember: the first design is rarely the simplest; low information content is achieved by exploring numerous alternatives.
- Consider material substitutions and design changes that facilitate the use of cost effective and time effective net shape manufacturing processes (e.g., plastic injection molding, powder metallurgy, etc.).

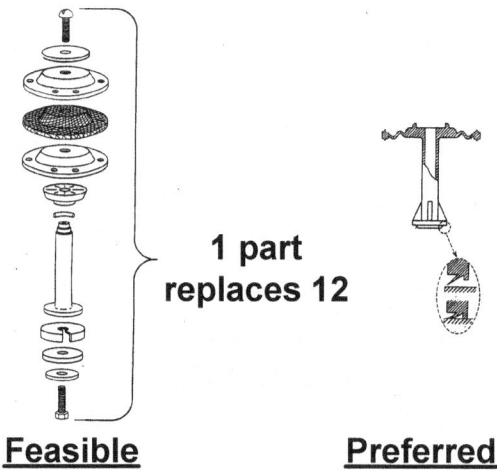

(Source: DuPont Hytrel Advertisement)

Design Hybrid Parts: Information content is reduced by shifting difficult assembly operations to the piece-part level. *Hybrid parts* are a form of integral design that allows material properties to be different for reasons of electrical conduction, strength, wear, etc. Care must be taken to ensure that increased tooling complexity, cycle times, and difficult material recycling don't offset simplification gains. A plastic molding in which threaded metal bushings have been inserted during the molding process is a common example of a hybrid part. Joining a shaft and gear as part of a die casting process is another.

Illustrative Example 4.1: Sometimes, when large part counts are unavoidable, the need to handle and assemble numerous parts can be avoided by postponing differentiation as is done during manufacture of a multi-contact strip shown below. The multi-contact strip is fabricated as a single metal stamping. The stamping is then insert molded into a plastic base part. The final step is to create the multi-contact assembly by cutting the connecting piece away.

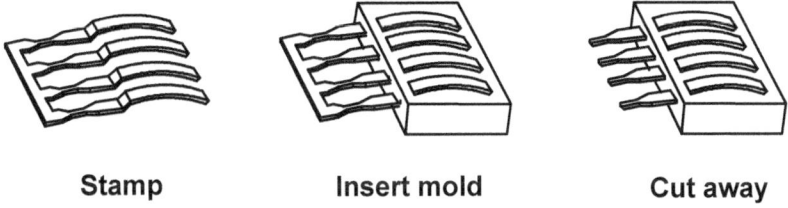

Stamp   **Insert mold**   **Cut away**

WARNING: Always be sensitive to the fact that functions may become coupled or interact in undesirable ways. (See illustrative example 3.1 on page 36.)

WARNING: Always be on guard against reducing part count too far. The alternative designs below trade information content of precision machining against that of adjusting and securing a bracket. Less parts is usually best, but not always. In this case, a knowledgeable manufacturing engineer can resolve the question.

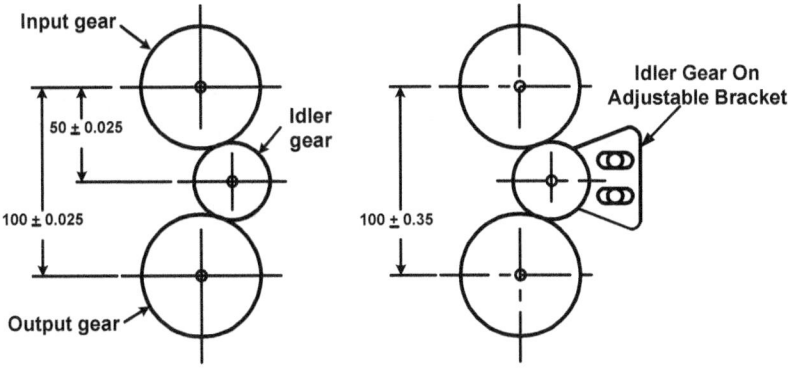

Standardize Part Designs: Standardization reduces information content by eliminating options and reducing the information content of the options that remain. Parts can be categorized as common parts, unique parts, and off-the-shelf parts. *Common parts* are in-house parts that are used across many designs and products. *Unique parts* provide one-of-a-kind functions and are usually exclusive to a single product or design. The piston of an internal combustion engine is a common part while the engine block is unique.

Standardize Common Parts: Design so that the same common part or component can be used interchangeably in different subassemblies, products, and applications. A catalog of *repeat parts* avoids the need to design and manufacture different versions of the same part. *Building block parts* allow the manufacture of different product variants simply by using different building block combinations.

Standardize Unique Parts: Because of their inherent nature, it is unlikely that unique parts themselves can be standardized. What is possible, however, is to group unique parts into part families and use the intrinsic similarities of each part family as a basis for standardization. The following example illustrates this approach.

Illustrative Example 4.2: Internal combustion engine blocks are usually sand cast from cast iron or aluminum and then finish machined on a large dedicated transfer line. Such transfer lines often take years to design, install, and prove out, they are extremely expensive and, because each engine block is unique, they become obsolete when production of the particular engine block they are designed to machine ends. The flexible manufacturing system (FMS) production line shown below avoids these drawbacks.

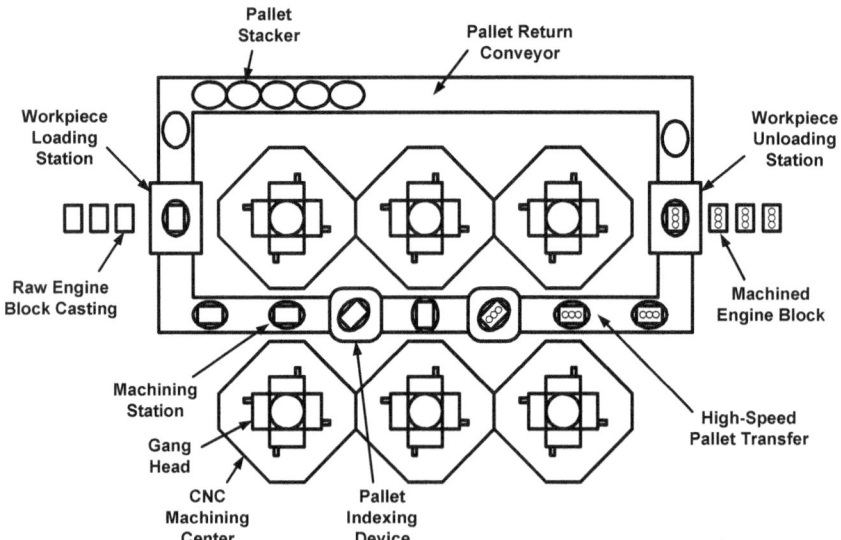

Each machining station of the FMS is supported by two fully programmable multi-axis CNC machining centers. To compensate for slower cycle times and allow simultaneous production of two or more different engine block designs, multiple identical FMS lines are operated in parallel. Each FMS line can be quickly and easily converted to production of a new engine block design simply by downloading appropriate NC programs. A new design is introduced into production by initially converting and proving out one or two FMS lines. As demand for the new engine builds and demand for the current design declines, more and more of the FMS lines are changed over to the new design. Because the FMS line is flexible, it is never obsolete.

Three design rules facilitate this FMS production solution: (1) the engine block envelop dimensions must be within prescribed maximums, (2) all machined features must be obtainable using three orthogonal machining axes, and (3) to reduce cycle time, holes and other features should be spaced to allow multiple simultaneous machining operations. By conforming to these simple design rules, each unique engine block design is standardized because each can be finish machined on the same FMS line. In essence, engine block design, production system, and production quantity are "decoupled", i.e., their independence is maintained.

Use Standard Off-the-Shelf Components: Information content is shifted to suppliers. Reliability is proven. Flexibility is improved because components can be purchased in any volume. Lead-time is generally short and up front investment is low.

Rationalize Options: Reduce the number of standard options to be used in future designs by understanding why some current options are popular and why unpopular options are sometimes used. Use this understanding to rationalize many options to a few "best" options.

Illustrative Example 4.3: Before rationalization, a company was purchasing 359 different fasteners. The number of choices after rationalization was reduced to 44. A comparison of fastener count and size distribution "before" and "after" rationalization is shown below.

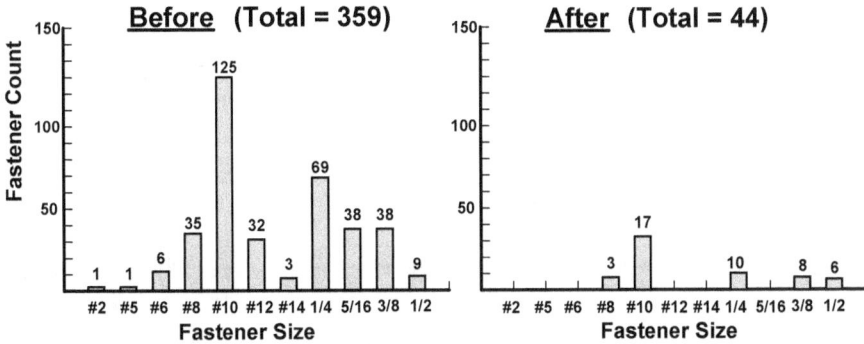

4.3 AVOID SEPARATE FASTENERS

Eliminate fasteners by integrating their function into mating parts. Fasteners themselves may be relatively inexpensive, but their installation is costly and ripe with quality risk. If all fasteners cannot be eliminated, design to reduce the fastener count. Consider "snap together" or interlocking designs. Reduce fastener types, head styles, drive styles, sizes, lengths, etc. When possible, identify standard options. Seek to reduce the number of standard options used in future designs.

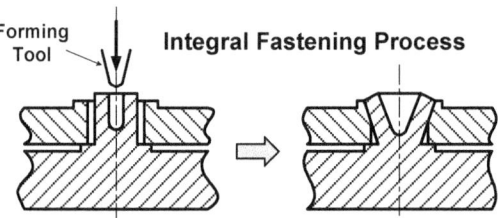

Use Snap-Fits: "Snap together" designs are easier to assemble than all other joining methods. Strength comes from mechanical interlocking and friction, so they can be very strong. Once assembled, snap-fits are in a low state of potential energy, which makes them vibration proof and resistant to creep. Snap assembly eliminates the need for special tooling and special operator training and can be tailored for permanent and non-permanent joining.

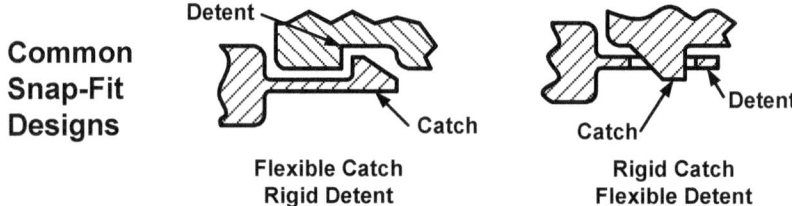

Minimize Fastener Variety: When separate fasteners must be used, limit sizes and styles.

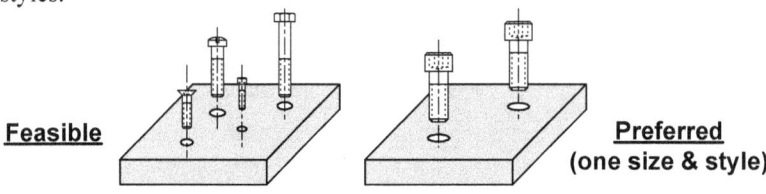

Design for One-Screw Assembly: Often, one screw is sufficient. Use molded features to augment holding strength.

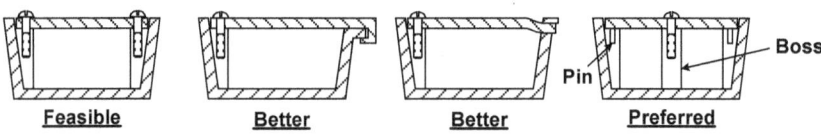

4.4 DESIGN PARTS FOR EASY FABRICATION

A component is easy to fabricate when tooling and process information content is minimized. Design so standardized and rationalized tools (drills, milling cutters, etc.) can be used and so die and mold cavities and punches are easy to machine. Minimize the number of individual processing steps, ensure that the design specification is well within the process capability, and select parameter values to minimize cycle time and cost while also maximizing yield.

Understand Geometry/Process/Material Interactions: Ease of fabrication requires compatibility between geometry, material and process capability.

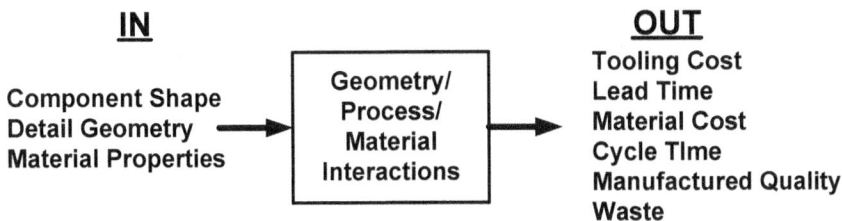

Reduce Processing Surfaces: Design so that the number of surfaces that must be processed is minimized. Try to design parts for orthogonal and/or parallel fabrication since most fabrication equipment is designed using perpendicular and/or parallel constructs. Design so processing can be completed on one surface before moving to the next.

Use Near Net Shape Processes: Near net shape manufacturing processes such as plastic injection molding, casting, powder metallurgy, and extrusion (shown below) produce minimal waste and often require minimal secondary processing (machining, deburring, etc.). They should be used whenever production quantities and development budgets allow.

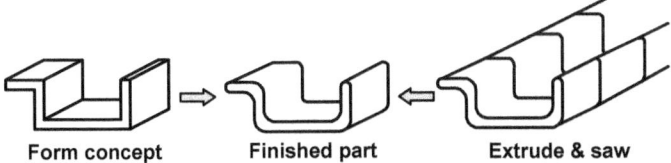

Form concept   Finished part   Extrude & saw

Avoid Secondary Processing: Finish machining, thermal and mechanical coating (painting), surface preparation, and so forth, require special setups, special facilities and equipment, and material handling systems. More costly material alternatives that avoid secondary processing often result in lower total cost because of reduced information content of the system as a whole.

Avoid Hard-to-Control Joining Processes: Avoid joining processes (1) that require excessive force, time, effort, part tolerances; (2) that require special tooling, equipment, worker skill; and/or (3) where joint integrity, quality, reliability is hard to detect and verify. Press-fit assembly, adhesive bonding, and welding, brazing, and soldering should be avoided when possible.

4.5 DESIGN PARTS FOR EASY ASSEMBLY

Processing steps will be decreased by simplifying assembly motion and reducing the number of different surfaces to which components are assembled.

Reduce Processing Surfaces: Ideally, components should only be added to one surface. When more than one surface is involved, design so that all assembly can be completed on each surface before moving to the next.

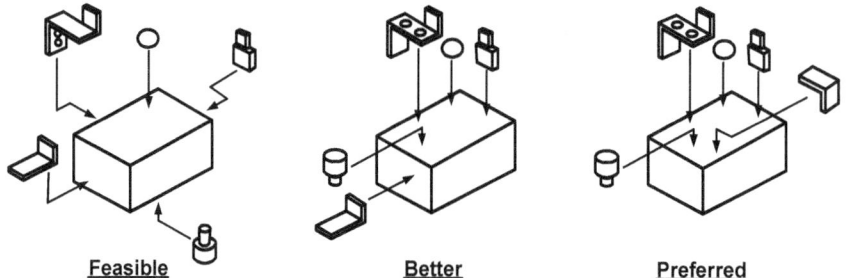

Minimize Assembly Directions and Motion: Design so that insertion motion is as simple as possible, preferably from one direction without changing direction. The ideal assembly direction is top-down because gravity assists the assembly process.

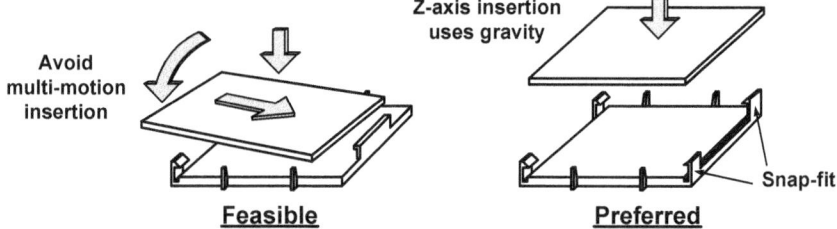

Eliminate the Need for Reorientation: Imagine that all parts are added top-down. Design so that this is possible without the need to reorient the partially completed assembly. Any required reorientation or manipulation of parts already assembled increases information content while adding no value to the assembly.

Design Assemblies as Layered Z-Axis Stacks:

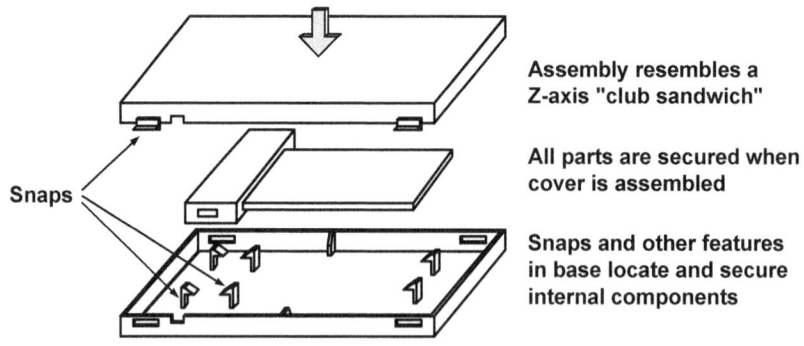

Assembly resembles a Z-axis "club sandwich"

All parts are secured when cover is assembled

Snaps and other features in base locate and secure internal components

4.6 DESIGN PARTS FOR EASY INSERTION

Information content involved in adding components to the partially built-up assembly is substantially reduced when (1) all components have high manufactured quality, i.e., predictable and controlled variation, (2) features such as generous radii, tapers, leads, and chamfers are provided to align, guide, and help center the part, (3) access, clearance, and vision are unrestricted, (4) the base component is rigid and accurately located, and (5) assembly tooling and automation, if used, has selective compliance to facilitate insertion alignment and minimization of insertion force.

Provide Generous Chamfers, Tapers, and Radii:

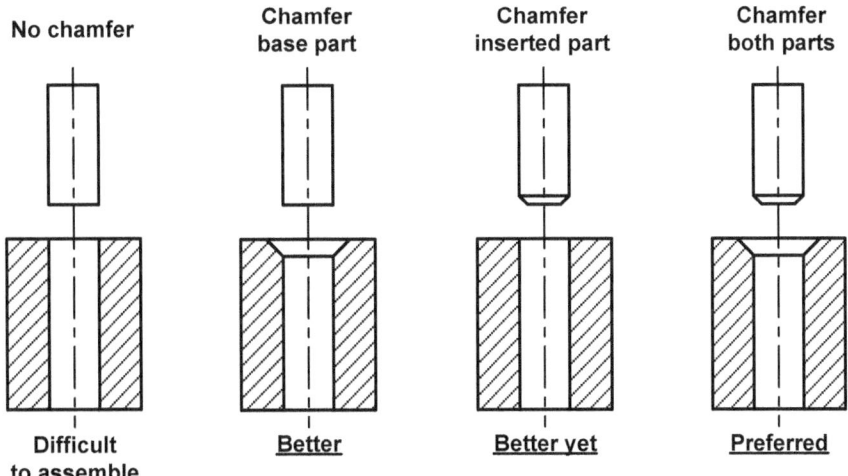

Insure adequate Access, Clearance, and Unrestricted Vision: Adequate access, clearance and unrestricted vision are essential for easy insertion. This guideline seems so obvious that it is often overlooked. As a result, access problems are discovered in production when they are difficult to correct. To avoid problems, anticipate possible difficulties by visualizing the assembly process early when the design is still relatively easy to change.

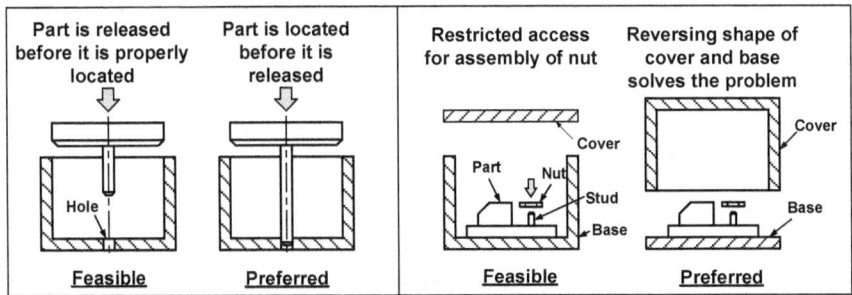

Allow Adequate Tooling Clearance: Adequate clearance permits the use of standard assembly tooling and readily available tools.

4.7 DESIGN PARTS TO BE SELF-ASSEMBLING

Design parts to do the job of assembly themselves rather than relying on assembly workers and fixtures. Design parts so that as they are added to the build, they are (1) correctly located and oriented by features on mating parts, (2) they do not need to be secured immediately, (3) they do not need to be held in place by external means before securing, and (4) no adjustment is required.

Design Parts To Be Self-Locating and Self-Aligning:

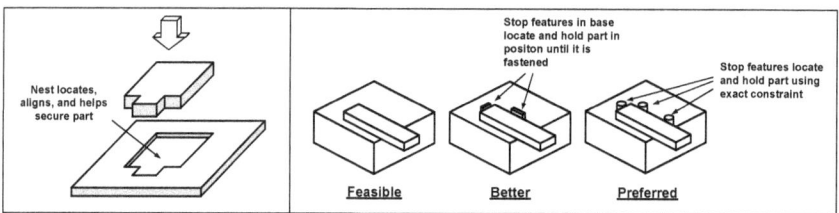

Provide Locators On Mating Parts: Provide notches, holes, pins, guides, or other features that align and locate the part. When possible, space locators far apart and use exact constraint.

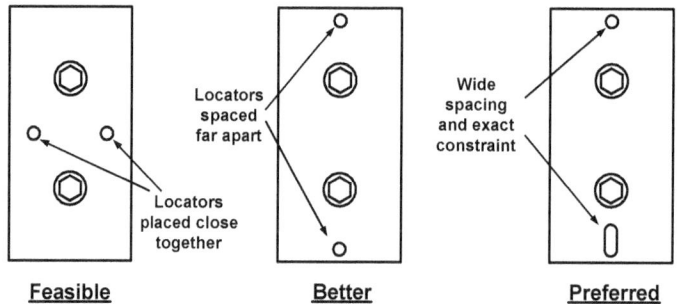

Avoid Adjustments: Adjustments are expensive, are a continual source of problems, and cause customer dissatisfaction. Avoiding adjustments reduces information content and quality risk. It is best if critical dimensions and orientations are incorporated into a single part. When this is not possible, use self-aligning and self-adjusting features, avoid multi-hand coordinated procedures, avoid over-constraint, use designed-in compliance, and never rely on subjective judgments (e.g., "fit at assembly").

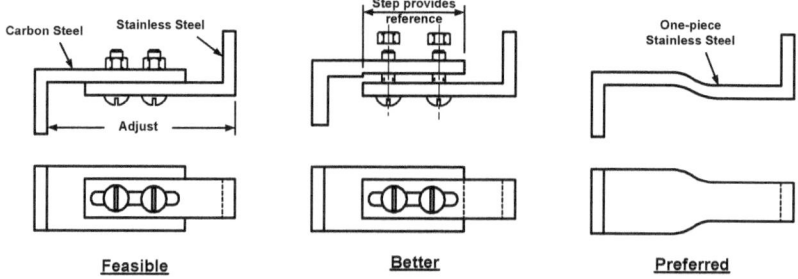

4.8 DESIGN PARTS FOR EASY HANDLING

Handling is the process of grasping, transporting, and orienting components. Parts that are easy to handle simplify automation and reduce quality risk.

Design Parts To Be Symmetrical: As orientations increase, equipment costs grow, quality risk increases, feed rate drops, and cycle time slows.

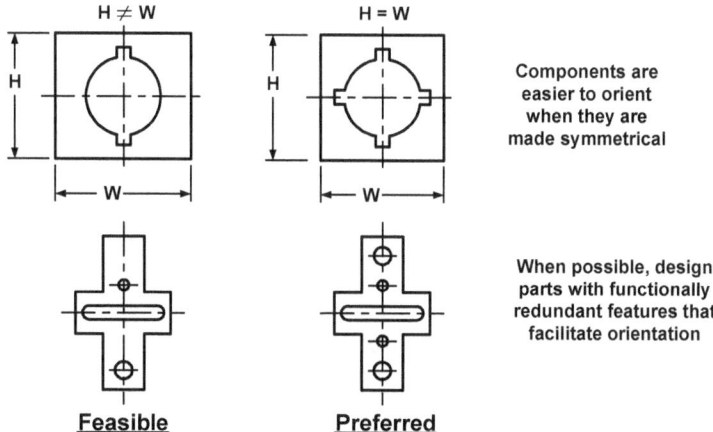

Components are easier to orient when they are made symmetrical

When possible, design parts with functionally redundant features that facilitate orientation

Make Asymmetry Easy To Recognize: If a part cannot be symmetrical, then make it visually and physically asymmetrical to avoid incorrect orientation.

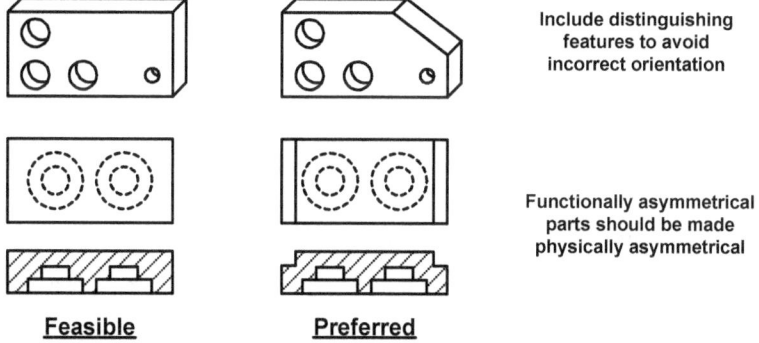

Include distinguishing features to avoid incorrect orientation

Functionally asymmetrical parts should be made physically asymmetrical

Avoid Handling Impediments: Avoid part features that can nest and tangle, that are sharp or slippery, that are fragile, or that are flexible.

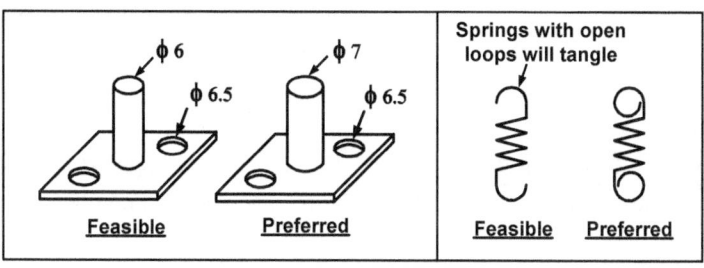

Springs with open loops will tangle

4.9 ERROR PROOF THE DESIGN

Information content is reduced when parts are designed so that they cannot be fabricated, handled, installed, or identified incorrectly.

Make Incorrect Assembly Impossible:

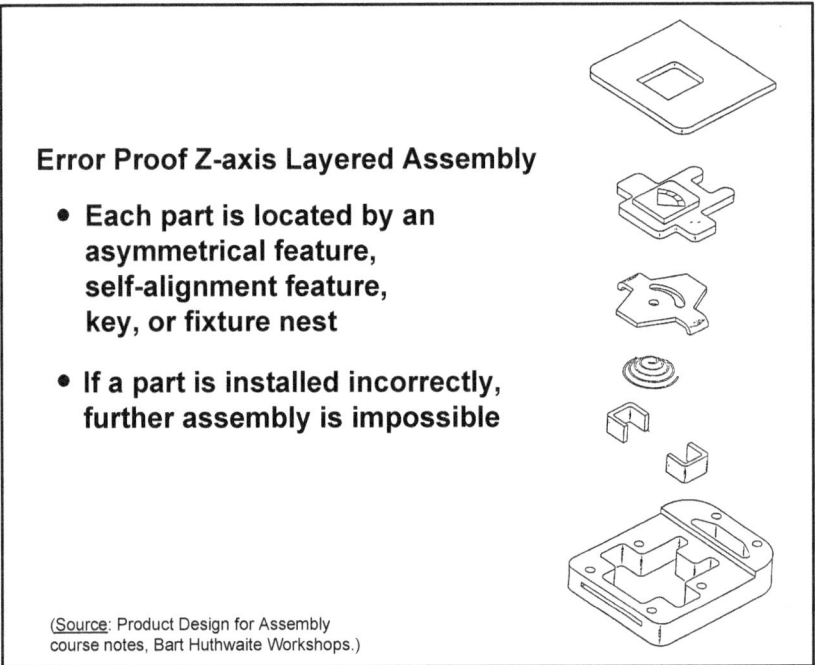

Error Proof Z-axis Layered Assembly

- **Each part is located by an asymmetrical feature, self-alignment feature, key, or fixture nest**

- **If a part is installed incorrectly, further assembly is impossible**

(Source: Product Design for Assembly course notes, Bart Huthwaite Workshops.)

- Provide obstructions that will not allow incorrect assembly.

- Make mating features asymmetrical.

- Make parts symmetrical so assembly orientation is unimportant.

- Make subsequent assembly impossible if a part is incorrectly installed.

- If necessary, provide clues such as matching arrows or colors. Note that this is less desirable that ensuring that incorrect assembly is impossible.

- Provide keys and other features on flexible parts such as gaskets to prevent incorrect installation. When possible, avoid flexible components since these can almost always be incorrectly installed.

Avoid Worker Judgment: Avoid designs that require workers to make decisions or judgments. Never rely on instructions such as "fit at assembly".

Mark Parts for Easy Recognition: Conspicuously color coded parts are more easily identified than parts that are identified by reading a part number.

4.10 ELIMINATE RANDOMNESS

Randomness and uncertainty increase unpredictability as well as information content and quality risk. Every effort should be made to eliminate randomness and when that is not possible, design to avoid undesirable effects.

Strive for Predictability: Randomness often masks a lack of clarity in the design. Be on constant lookout. Do your utmost to explain all behaviors and phenomena associated with the design. If a behavior can't be predicted or explained in simple and logical terms, then look for undesirable interactions.

Avoid Flexible Parts: The orientation, shape, and position of flexible, slack, and/or deflecting parts are uncertain and hard-to-control. If flexure is necessary, then strive to ensure that the part shape, etc. is as predictable as possible.

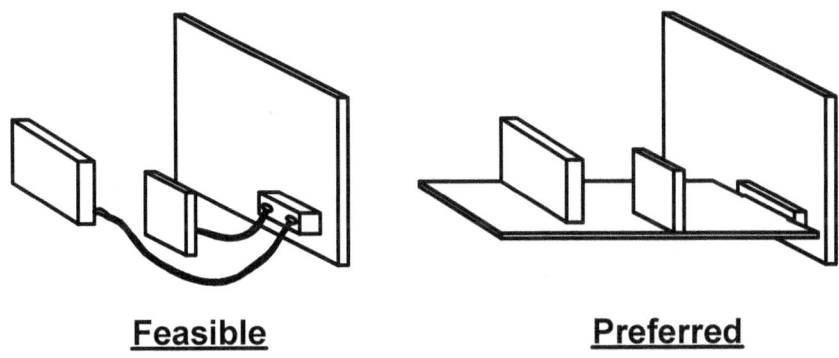

Feasible   **Preferred**

Restrain Loose Wires and Wire Harnesses: Provide features that orient, locate, and restrain wire harnesses and other components that tend to be randomly located. Plan the wiring layout early in the design to (1) minimize the amount of wire used; (2) constrain otherwise arbitrary decisions regarding location of connectors on "black boxes" and subassemblies; and (3) facilitate logical placement of features that orient, locate, and restrain loose wires and wire harnesses.

4.11 PART DECOMPOSITION IMPROVEMENT METHOD

This method is a systematic procedure for optimizing the part decomposition of an existing or proposed design. A design-analyze-redesign procedure is used to first analyze an existing or proposed design and then, using the insights gained from the analysis, develop redesign alternatives that optimize the part decomposition in light of the rules of good design. The method is performed in four steps.

1. **Gather Information**: Obtain the best available information about the design.
    *   Engineering drawings
    *   An existing version of the design or a working prototype

2. **Analyze**: Determine part decomposition improvement opportunities.
    *   Take the design apart (or imagine how this might be done). If the design contains sub-assemblies, treat these, at first, as "parts" and then analyze them later as assemblies.
    *   Begin reassembling the design in the reverse order from which it was disassembled. Imagine that the build can be reoriented so that each part is added top-down using one-hand and keep track of each reorientation. As each part is added to the assembly and regardless of practical or functional limitations, answer each of the following questions:
        (1)   Does the part move relative to other parts?
        (2)   Must the part, for good reasons, be made of a different material?
        (3)   Does the part need to be separate for assembly or service?
    *   If the part receives an answer of "Yes" to _any_ of these questions, then the part is a theoretical part and _probably_ cannot be eliminated. If the answer to all three questions is "No", then the part is a candidate for elimination (CFE).
    *   As each part is added to the build, analyze it for opportunities to eliminate friction and information content by avoiding reorientations and improving ease of handling and assembly.
    *   Record improvement opportunities using copies of the "Part Decomposition Improvement Worksheet" shown on the next page. Alternatively, use an equivalent computer "spreadsheet".

3. **Redesign**: Implement improvement opportunities identified in Step 2. To maximize the possibility of identifying a "best" redesign, develop several alternative redesigns that range from simple minor changes to speculative (and risky), but desirable "far-out" ideas. Eliminate friction and information content. Target the part decomposition optimization goals.

4. **Winnow, Refine, Optimize**: Develop improved redesign concepts by eliminating undesirable features and combining desirable features.

PART DECOMPOSITION IMPROVEMENT METHOD WORKSHEET

Name of Design _____   Sheet _____ of _____

Part Name	Qty	Motion	Material	Assembly/ Service	CFE	Notes

$$\sum Parts = \underline{\hspace{1cm}} \qquad \sum CFE = \underline{\hspace{1cm}}$$

$$Count\ Efficiency = \frac{\sum Parts - \sum CFE}{\sum Parts} \times 100\ \%$$

Illustrative Example 4.4: Improve the part decomposition of the vacuum cleaner attachment shown using the part decomposition improvement method.

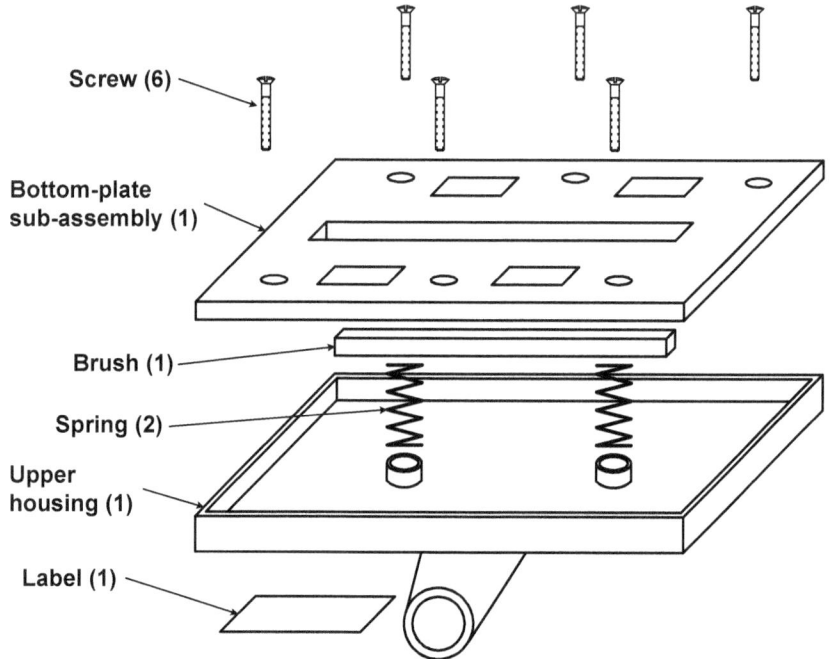

Six screws hold the bottom plate subassembly to the upper housing. The bottom plate subassembly, which consists of the bottom plate together with several pressed in pin-mounted rollers that facilitate movement of the attachment along the floor surface being vacuumed, is treated initially as a single part. An asymmetrical nylon brush is sandwiched between the bottom plate sub-assembly and the upper housing. The brush extends through a slot in the bottom plate and is held in contact with the floor by two coil springs which fit into circular nests molded in the upper housing. The vacuum attachment is disassembled by removing (1) six screws, (2) one bottom plate assembly, (3) one brush, (4) two springs, and (5) one label. The upper housing is the last remaining part. This part is designated as the "base" part or "starting" part.

Analysis: Starting with the upper housing (the base part), reassemble in the reverse order. Analyze each part as it is added to the build by completing one line on the worksheet after it has been added "top-down" to the build using one hand. The part count is recorded in the quantity column and any assembly difficulties are noted. The three critical questions are asked of each part and if the answer is "no" for all three questions, the part is a candidate for elimination (CFE) and the part count that is recorded in the quantity column is also recorded in the CFE column. Separate operations, such as the reorientation required after the label is added, are recorded on a separate line of the work sheet and are always CFE's. The worksheet and analysis for the vacuum attachment is shown as follows.

PART DECOMPOSITION IMPROVEMENT METHOD WORKSHEET

Name of Design _____Vacuum Attachment_____   **Sheet** __1__ **of** __1___

Part Name	Qty	Motion	Material	Assembly/ Service	CFE	Notes
Upper Housing	1	No	No	Yes	0	
Label	1	No	No	No	1	Sticky; hard to position
Reorientation	1	—	—	—	1	
Spring	2	No	No	No	2	Tangle; hard to handle
Brush	1	Yes	No	No	0	Orientation not obvious; holding required
Bottom Plate Sub-assembly	1	No	No	Yes	0	Hard to guide brush through slot
Screw	6	No	No	No	6	

$$\sum Parts = \underline{\quad 13 \quad} \qquad \sum CFE = \underline{\quad 10 \quad}$$

$$Count\ Efficiency = \frac{\sum Parts - \sum CFE}{\sum Parts} = \frac{13-10}{13} \times 100 = 23\%$$

- Since the base part is the first part in the assembly, the motion and material questions both receive "no" answers. The assembly/service question is a "yes" however, because a base part is needed to begin the assembly.
- As noted, the label is difficult to handle because it is sticky on one side. It is hard to position because it is flexible.
- The reorientation is necessary so that the springs can be added "top-down". The method assumes that the build remains in tact regardless of orientation.
- The coil springs are difficult to handle because they tend to tangle and nest. A "no" answer is assigned to the motion question because one end of the spring is stationary.
- The brush is difficult to handle because it is asymmetric and the correct orientation is not obvious. It is difficult to insert because the springs need to be held in position.
- The bottom plate is difficult to insert because there is no taper or other feature to guide the brush bristles through the slot. The assembly/service question is answered "yes" for two reasons: (1) the part is a sub-assembly and (2) it must be separate to allow assembly of the springs and brush.
- Separate fasteners will always receive "no" answers to the critical questions and are always CFE's.

Re-Design: The part decomposition is improved by seeking creative ways to harvest the improvement opportunities identified in the analysis step. The goal is to eliminate CFE's and to design the parts that remain to be easy to assemble and manufacture. To fully explore all improvement possibilities, it is recommended that the redesign proposals developed include a "practical" redesign, a "stretch" redesign, and a "radical" redesign. A practical redesign is one that can be implemented fairly easily without excessive tooling costs or development effort and with no potential change to design function, performance, or capacity. A stretch redesign, on the other hand, could involve significant development effort and cost and may also impact performance to some extent. A radical redesign is a stretch redesign that involves a major change such as switching to a totally different physical concept or to a different material or manufacturing process. To illustrate, we propose the following vacuum cleaner attachment redesigns.

Concept A, Practical: (1) Eliminate the label by integrating it into the upper housing either as a "hot stamping" or as a molded-in feature. (2) Replace the screws with snap fittings. This can be done by making relatively simple changes to existing tooling. (3) Mold-in a tapered guiding feature on the bottom plate to orient the brush and guide the bristles through the slot. Again, this requires simple modification of the existing tool. (4) Provide snap-fitting in the upper housing to retain the brush thereby avoiding the current instability and holding requirement. (5) Assemble as a "Z-axis" stack without the need to reorient the build.

Concept B, Stretch: Same as concept A plus eliminate the separate coil springs by integrating the spring function into the brush. One approach might be to incorporate cantilever beams on the brush as shown below.

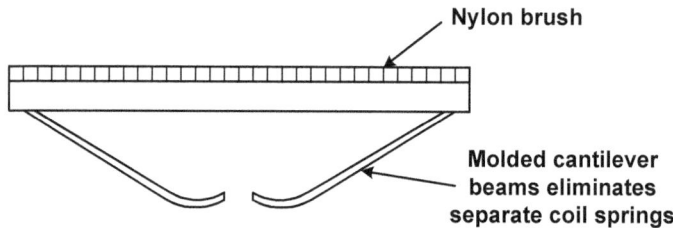

Nylon brush

Molded cantilever beams eliminates separate coil springs

Concept C, Radical: (1) Combine the bottom plate and upper housing into a single part. (2) Integrate the brush and coil springs into a single part that snap-fits into the upper housing and that is exactly constrained by molded-in features.

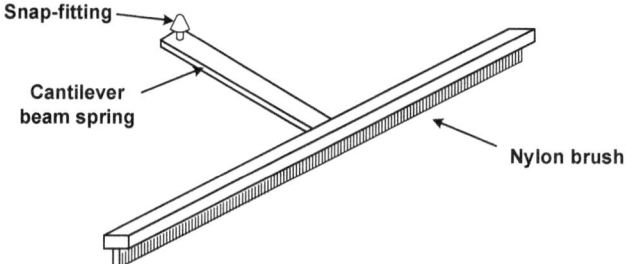

Snap-fitting

Cantilever beam spring

Nylon brush

Discussion: Redesign concepts A, B, and C are compared as shown in the following table.

Redesign Concept	Part Count	Parts Eliminated	Sub-Ass'y Eliminated	Count Efficiency
Original Design	12	---	---	23 %
Concept A	5	7	0	60 %
Concept B	4	8	0	100 %
Concept C	2	10	1	100 %

This comparison illustrates that significant part count reduction is possible using a practical redesign (concept A). It also shows that further part count reduction tends to be accompanied with increased tooling cost and implementation complexity (concepts B and C). At the same time, reduction in information content can be remarkable and well worth the effort. Concept C, for example, eliminates all of the information content associated with the bottom plate sub-assembly and replaces a 12-part design (not counting the parts in the sub-assembly) with one having just two separate parts. It should be noted that, in creating concept C, the team had to reevaluate the "yes" answer to the "assembly/service" question for the bottom plate assembly. This illustrates a subtle pitfall associated with the three critical questions. Often a "yes" to one of the questions is the result of the particular part decomposition being considered and can be converted into a "no" by a different decomposition approach as illustrated by concept C. Therefore, answering "yes" to a critical question should always be carefully considered and all "yes" answers should be challenged during the redesign step. This is especially true for the "assembly/service" question. It should also be noted that there are many possible metrics that can be used to measure the "quality" of the design with respect to friction. The "count efficiency", which compares theoretical part and separate operation count to the actual count, acts as a simple to calculate measure of the information content. As illustrated by concepts B and C, however, it is also an imperfect measure because it depends on how the three critical questions are answered. Therefore, in practice, it is important to focus on reducing friction in the design rather than on improving metrics such as the count efficiency.

***Winnow, Refine, and* Optimize:** In the "winnow, refine, and optimize" step, the performance and cost implications of each redesign proposal are evaluated in light of customer needs and development time and budget constraints. "Hybrid" concepts that combine the best features of each design proposal are also developed and evaluated. For example, a "hybrid" concept D can be created by modifying concept B to use the spring/brush design from concept C.

4.12 ADVICE, INSIGHTS, AND CAVEATS

Iterate to minimize information content. It's been said before, but bears repeating. The first part decomposition is seldom the one having the least information content. It is best to propose a trial part decomposition that maintains independence of functional requirements, analyze it in light of the rules of good design, and then use the insights gained to redesign for reduced information content. Use the Part Decomposition Improvement Method (pages 58) to provide a disciplined and systematic approach.

Cost-reduce current designs. Almost all existing designs can be cost reduced by optimizing the part decomposition. Use the Part Decomposition Improvement Method (page 58) to provide a disciplined and systematic approach.

Challenge theoretical parts. When a part receives the answer "Yes" to one of the critical questions, the natural assumption is that the part is a theoretical part and cannot be eliminated. This is not always the case. For example, parts that undergo small relative motions can sometimes be eliminated by using a "flexure" or "living hinge". Often it is the physical concept or the part decomposition itself that is forcing the "Yes" answer. Look for an alternative concept that avoids the "Yes". Always challenge theoretical parts.

Create integral designs by integrating function. Combine functions when possible. For example, design a part to act as an electrical conductor and also as a structural member. Use an electronics chassis to carry load, dissipate heat, and act as an electrical ground. *WARNING*: do not allow functions to become coupled or to interact in undesirable ways.

Design for one-hand assembly. Design so parts can be grasped, oriented, positioned, and added to the build using **one hand** without special tools.

Maximize commonality. Commonality should be maximized at all levels of the design. Use common hardware such as washers and ball bearings, use common geometric features and dimensions on all designed parts, use common manufacturing processes and tooling to produce the parts. To maximize commonality, planning must begin in the very early stages of the design. Define a short list of acceptable hardware (e.g., fasteners) and common components (e.g., ball bearings) and use these exclusively in the design. Similarly, define a minimum set of permissible readily available tools and easy to perform service operations. Common mechanical and electrical interfaces simplify manufacture, assembly, maintenance, and service. Plan the interfaces early to guide all subsequent "black box" design and system integration decisions.

Get suppliers to do as much as possible. Have the supplier capture and maintain part orientation. Material handling schemes that assist and maintain orientation include palletized trays, magazines, tube feeders, part strips, and kitting.

SECTION 5

COORDINATED DESIGN

In "coordinated design", the part decomposition is optimized by designing the geometrical layout and assembly concept as a coordinated system. The geometric layout defines the location and spatial arrangement of the individual parts that make up the part decomposition; the assembly concept defines the assembly sequence and structure. In addition to providing required functionality, the coordinated geometric layout and assembly concept determine how the design accommodates market and technological change, enables customization, adapts to standardization, and lends itself to ease of manufacture, assembly, and life-cycle support. In this section, we apply the Design for Everything approach to coordinated design. We do this by grouping parts together into "chunks" that provide explicit advantages and then integrate the "chunks" together by coordinating the geometric layout with an envisioned assembly concept. The coordinated design is optimized by avoiding undesirable interactions and minimizing information content.

5.1 DEFINITIONS AND BASIC CONCEPTS

Functional Element: Individual operations and transformations that contribute to the overall performance and functionality of the design. Some typical functional elements include "enclose", "support", "separate", "control", "channel", and so forth. Functional elements embody the "how" of the design and are often main, critical, or auxiliary functional requirements. They are usually described in schematic form before they are reduced to specific technologies, components, or physical concepts (see Section 1.5).

Physical Element: Parts, components, and subassemblies that work together to provide the design functions. Some physical elements are dictated by the physical concept (e.g., piston), some by the function they provide (e.g., electric motor), and some by the assembly concept (e.g., base component, frame, etc.).

Chunk: A collection of physical elements grouped together in a logical way to form a major building block of the design. Chunks can take many different forms. An individual part such as a large casting or housing can be a chunk. A standard off-the-shelf component such as a pump can be a chunk. A major sub-assembly or separate module can be a chunk. A group of parts such as gears in a speed reducer can form a chunk. At one extreme, the whole design might be lumped together as one chunk while at the other extreme, each individual part might be treated as a chunk. Chunks are defined internally within the firm and need not have any meaning or importance to the end user or other external customers. For good design, it is important that each chunk provide a well defined benefit for customers and/or clearly understood advantage for the firm.

Module: A chunk that is totally self-contained and that is equipped with standardized interfaces that allow it to be used interchangeably in different designs or within a particular design or application. Interchangeable camera lenses are an example. A battery pack that can be used interchangeably with a family of power tools is another example. Designs that allow larger systems to be built up by using multiple smaller standardized repeat physical elements (e.g., drawers used in a furniture cabinet) are yet another example.

Standard Component: A physical element and/or chunk that is used interchangeably in a variety of different designs and applications. Standard "off-the-shelf" physical elements such as light bulbs, electrical connectors, and mechanical fasteners, are categorized as *external* standard components. *Internal* standard components, on the other hand, are standardized building block physical elements, chunks, or modules that are unique to the firm and its family of products.

Designed Component: A unique physical element such as a part, component, or subassembly that is designed in its entirety to meet specified needs of the design. Choosing between designed and standard physical elements can be critical. For example, the choice of designing a special electric motor, optimized for performance and weight, as opposed to purchasing a standard off-the-shelf motor, can have far reaching performance, cost, quality, and timing consequences.

Part Decomposition Schematic: A part decomposition schematic is a diagram representing the constituent elements of the design. Some of the elements in the schematic are physical elements, some are defined physical components such as external or internal standard components, and some are functional elements that remain described only functionally. The schematic should reflect the best understanding of the state of the design, but it does not have to contain every detail. Ulrich and Eppinger [4] suggest that the total number of elements shown on the schematic be limited to 30 elements or less.

Illustrative Example 5.1: A hand held electric power drill is to be designed. The design specification requires that the internal standard trigger and speed control module that is used across many of the firm's products be used. It also requires that external standard parts include an off-the-shelf drill chuck and a purchased power cord set. Develop a suitable part decomposition schematic.

Overall Function Structure: As a starting point, we define the overall function structure for the design problem as shown below.

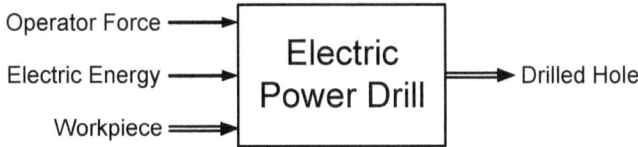

Possible Schematic: One possible part decomposition schematic for the hand held electric power drill is shown below. Note the presence of both functional elements (e.g., "channel force") and physical elements (e.g. "drill chuck"). Importantly, note that alternative schematics are possible. Note also that, as required by the problem statement, the internal (speed control module) and external (drill chuck, cord set) standard components are shown as separate physical elements.

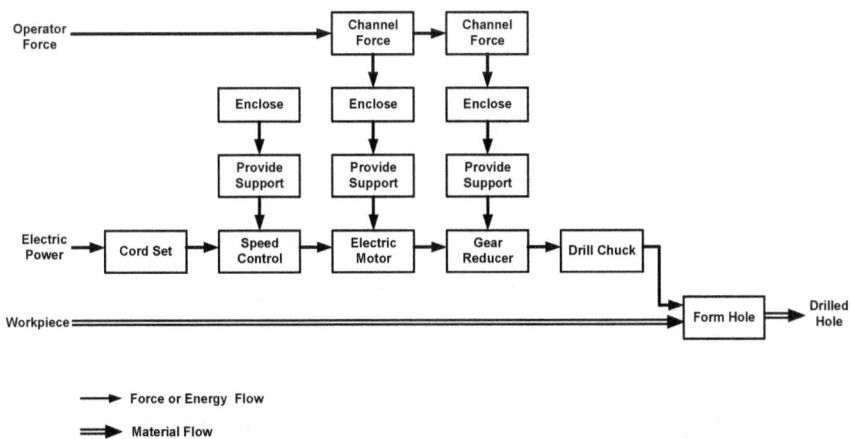

Geometric Layout: The spatial arrangement of physical elements and/or chunks that implement the physical concept while also satisfying aesthetic, user interface, manufacturing, and life-cycle support requirements and constraints. The geometric layout begins as a sketch or possibly as a solid computer model or 3-D physical model (e.g., cardboard or foam) in which the shape and size of components are approximate and/or represented by simple "placeholder" geometries. It is preliminary because, at most, only key dimensions and relationships have been specified; the actual size, shape, and detail features of the parts and possible chunking strategies are, as yet, either undefined or only partially defined. The preliminary layout is developed into a completed design by first deciding on how to divide the part decomposition into chunks and then developing a configuration design and parametric design for each chunk and designed component. *Configuration design* involves determining the size, shape, and detail features of the chunks and designed components; *parametric design* involves assigning specific material properties, dimensions, and tolerances. During configuration and parametric design, the preliminary layout changes and evolves as questions are answered and uncertainties resolved. The definitive geometric layout or final design is completed in the *detail design* stage where each detail of the layout is fully specified. The definitive layout contains the design information required to fabricate and assemble the components. In modern practice, the definitive layout is typically a computer solid model representation.

Illustrative Example 5.2: Disassembled views of electric drill products sold by two different manufactures are shown below. Each electric drill product is based on the part decomposition schematic shown in illustrative example 5.1, but the geometric layout and "chunking" of each differs. In Electric Drill A, the motor and housing (plastic injection molded base and cover) form a first chunk, the gear reducer forms a second chunk, and all remaining parts are each their own chunk. In Electric Drill B, the electric motor forms a chunk while all other physical elements are each their own chunk. Each layout offers advantages and disadvantages. The layout of Electric Drill A utilizes load paths that favor function. The layout of Electric Drill B, on the other hand, favors ease of assembly (see Section 4).

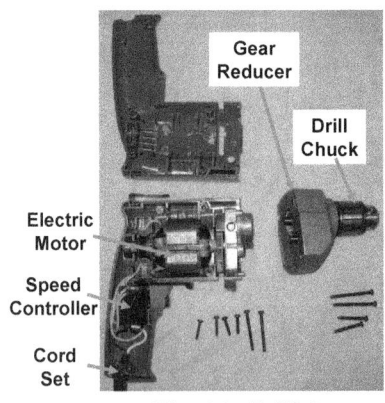

Electric Drill A

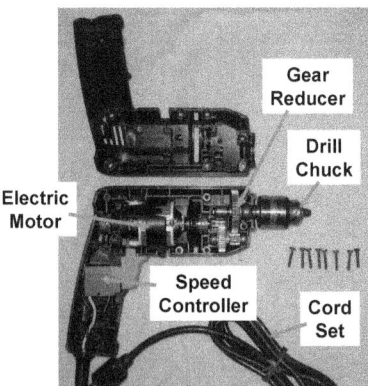

Electric Drill B

Material and Process Selection: An important part of any design is the selection of the basic material class (e.g., metal, plastic, wood) and type of process (e.g., machining, sheet metal forming, near net-shape processing) to be used in fabricating the major designed components. In many mature industries, material and process are determined or implied. For example, the exterior components of a typical passenger car are likely to be formed from sheet metal. In new product development, the options can be much broader. An enclosure for a small electronic device (think computer or cell phone) could be machined from solid metal, formed out of sheet metal, designed as an aluminum casting, or made of an engineered plastic using a variety of polymer processing methods. The actual choice depends on a number of factors such as production quantity, required ruggedness and shielding, and in-house manufacturing capability and expertise to cite just a few.

Assembly Structure: The way in which the chunks and physical elements are supported, oriented, located, joined, and integrated together to form the whole. Most assembly structures utilize a chassis or frame on which components are mounted. Frames can be oriented vertically (e.g., bicycle or motorcycle frame), horizontally (e.g., rail-type automotive chassis or electronics chassis), or designed as 3-D skeletons (e.g., space frame, unibody, card cage, etc.). Parts can be located, oriented, and joined to the frame or base part in different ways such as the use of nesting features or the use of a multitude of different joining processes. As discussed in Sections 4.5, 4.6, and 4.7, information content is reduced by eliminating the number of processing surfaces and insertion directions, designing for top-down z-axis assembly, and avoiding separate fasteners. In *stacked construction*, for example, parts are assembled top-down onto a base part, located by fixture nests, and held together by a cover or other top part. A well-thought-out assembly structure is often the secret to easy assembly.

Assembly Sequence: Starting with the base part, the assembly sequence is the sequential order in which the physical elements (parts, components, sub-assemblies, and chunks) are added to the partially completed assembly (often called the "build"). Ambiguity and information content are reduced by deciding on the assembly sequence early in the design and then designing the physical elements to be assembled accordingly.

Assembly Concept: The overall assembly scheme or plan employed for assembling the final product. The assembly concept defines the assembly structure and assembly sequence, together with how the physical elements are to be handled, inserted, retained, secured, checked for correct assembly, and tested for defects and acceptable performance. Developing an assembly concept as part of the geometrical layout helps ensure that assembly requirements and constraints are considered early. Undesirable interactions are avoided, information content is minimized, and potential problems are identified and corrected long before they become real problems on the manufacturing floor.

Assembly Process: Specification of the assembly method, the assembly line layout, allocation of work elements (line balancing), workstation layout, etc.

Illustrative Example 5.3: Consider an automobile design in which the physical elements have been assigned to three major chunks: the body-in-white (frame), the rear axle assembly, and the power train assembly. As shown below, a possible assembly process for this assembly concept is to (1) load the body-in-white onto the assembly line, (2) assemble interior and exterior components, (3) add the rear axle chunk and the power train chunk at the final two stations.

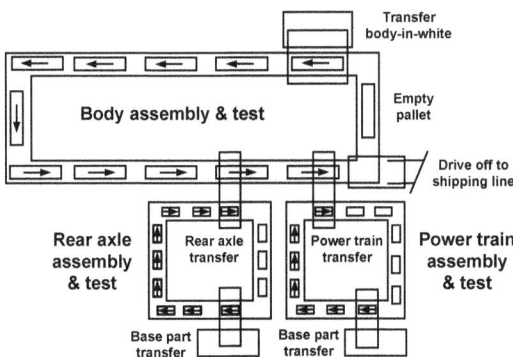

Assembly Methods: There are three basic methods of assembly: (1) manual assembly performed at a bench or on a transfer-line; (2) special-purpose transfer machine assembly (flexible or fixed automation) in which assemblies are either transferred by an indexing transfer device (rotary or in-line) or by a free-transfer non-synchronous device; and (3) robot assembly. In practice, assembly systems can be a combination of one or more of these methods.

Transfer Line: A transfer-line consists of multiple workstations arranged in sequence, with the partially completed assembly or "build" physically moved from station to station to complete the assembly process. Many variations are possible.

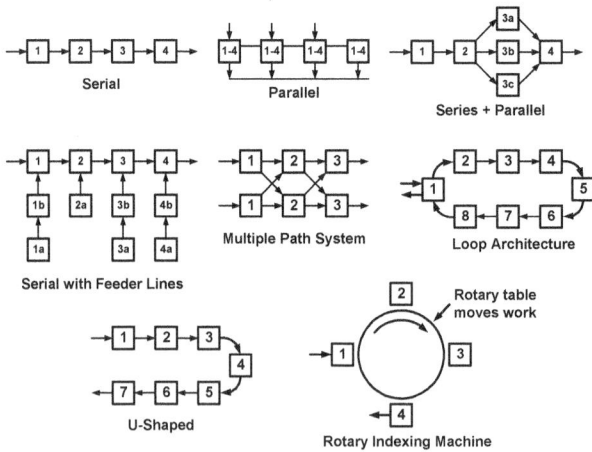

Design Imposed Assembly Constraints: The particular part decomposition selected imposes a variety of constraints on the assembly process.

1. *Work content*: the work content (sometimes called assembly content) is the time required to complete assembly of the design. It is established by the part decomposition and usually cannot be reduced without redesign.

2. *Minimum rational work element*: the smallest practical amount of work into which the work content can be divided. Often, different work elements will require different times, and when they are grouped into logical tasks and assigned to workstations, the task times will not be equal.

3. *Precedence constraints*: the part decomposition imposes restrictions on the order in which work elements can be performed. Some elements must be completed before others. A certain element that might be allocated to a workstation to achieve a better line balance cannot be added because it violates a precedence constraint.

4. *Minimum number of workstations*: The number of workstations required to accomplish the total work content in a specified production time (time available to assemble one unit in order to produce a required production quantity per minute, hour, day, or year). For a "series" transfer line having perfect line balance, the minimum number of workstations is,

$$Minimum\ number\ of\ workstations = \frac{Work\ Content}{Production\ Time}$$

The minimum number of workstations is an ideal whose achievement in practice is difficult for the following reasons:

- *Imperfect line balancing*: it is very difficult to divide the work equally among workstations; some stations will be assigned work that requires less than the production time.

- *Task time variability*: there is inherent variability in the time required to perform a given task; this is especially true on manual assembly lines.

- *Quality problems*: defective components and other quality problems will cause delays and rework that will add to the work load.

- *Repositioning time losses*: some time will be lost at each station because the build must be transferred from workstation to workstation. Therefore, the amount of time available at each workstation is less than the production time.

Imperfect line balancing, task time variability, and quality problems can often be reduced or eliminated by coordinated design, i.e., when the part decomposition and assembly process are designed as a coordinated system.

Illustrative Example 5.4: Because of the close coupling between geometric layout and assembly sequence, information content can be greatly reduced by designing the assembly sequence as part of the part decomposition optimization. Consider the "speed stick" product shown below.

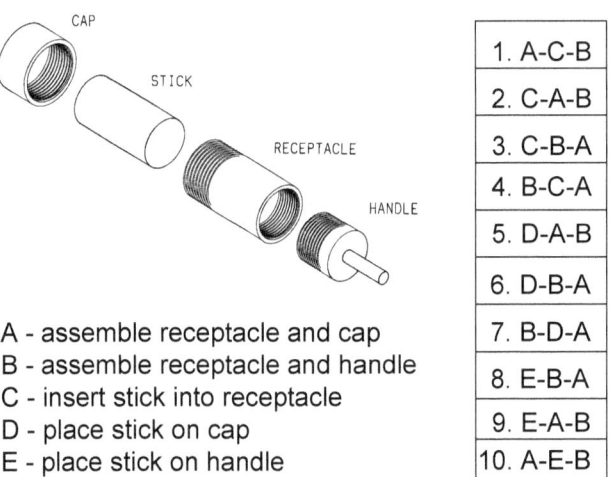

A - assemble receptacle and cap
B - assemble receptacle and handle
C - insert stick into receptacle
D - place stick on cap
E - place stick on handle

1. A-C-B
2. C-A-B
3. C-B-A
4. B-C-A
5. D-A-B
6. D-B-A
7. B-D-A
8. E-B-A
9. E-A-B
10. A-E-B

Traditional Approach: The design shown is released to manufacturing by design engineering. The assembly engineer decides on an assembly sequence and develops an assembly process accordingly. Although there are only four parts, amazingly, there are ten possible assembly sequences as shown. Any of these ten assembly sequences can be arbitrarily selected. The assembly engineer must select the best sequence for minimizing the number of workstations and distributing the assembly work as evenly as possible. Imagine the number of different assembly sequences that are possible in products with higher part counts such as appliances and automobiles and you begin to glimpse the assembly engineering challenge. You also begin to appreciate the wisdom of coordinated design.

Coordinated Design Approach: The team, including design and manufacturing engineers, develops the assembly sequence as part of the part decomposition optimization. This allows the team to optimize the design with respect to the assembly process. For example, suppose the first assembly sequence in the list (A-C-B) is selected. By knowing this assembly sequence, the work content can be minimized by (1) adding "flats" or other suitable features to the handle to make it easy to clamp in a vice and hold vertically for "top-down" assembly; (2) tapering the stick on the appropriate end to ease insertion into the receptacle; and (3) designing the cap for easy top-down threading onto the receptacle. Best of all, the assembly sequence is specified as part of the design thereby freeing the assembly engineer from the ambiguous chore of determining precedence constraints and rational work elements.

Illustrative Example 5.5: A company wishes to manufacture four versions of a product for a total of 15,000 units per year. Work content is estimated to be 45 minutes. The effective work time per shift is 420 minutes and the line is to be operated for 2 shifts per day for 240 days/year. Required production time is calculated to be 11.4 minutes after being reduced by 15% to compensate for uncertainty. Develop an appropriate assembly process (a) using the traditional approach and (b) using a coordinated design approach.

(a) Traditional Approach: The design is released to manufacturing. Assembly engineering analyzes the design and decides to use a sequentially arranged multiple workstation transfer line in which all assembly operations are performed manually by human workers. A work carrier is used to hold the base part during assembly. The base part travels through each workstation, where workers add parts and perform tasks that progressively build the product. A synchronous work transfer system moves work from station to station with a repositioning time of 1.0 minute. Based on constraints imposed by the model mix, assembly sequence, precedence, and repositioning time, an assembly transfer line having 6 workstations is developed. Actual cycle time for each workstation is shown below.

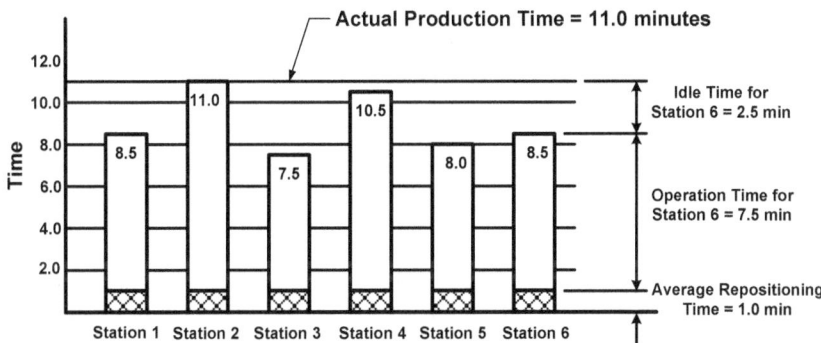

(b) Coordinated Design Approach: The team designs the product to be assembled in its entirety by one worker at one workstation. Parts are delivered using kits that establish orientation and provide the required part mix for each product model. Information content of the part decomposition and assembly sequence combination is minimized using the optimization guidelines of Section 4 to give a work content time of 24 minutes. This permits two parallel workstations, each having a cycle time of 12 minutes, to be used. In addition to the savings tabulated below, the coordinated design approach also reduces factory floor space, material handling complexity (and cost), and workstation cost.

	Traditional	Coordinated
Work Content	45.0 min	24.0 min
Minimum number of workstations	4	2
Actual number of workstations	6	2
Actual production time	11.0 min	12.0 min

5.2 GUIDELINES FOR COORDINATED DESIGN

The objective of coordinated design is to (1) develop a scheme by which the physical elements of the part decomposition are grouped together into logical "chunks" that provide explicit benefit to customers and/or advantages to the firm; (2) devise an assembly concept that will help guide and inform the design process; and (3) provide coordinated design features that facilitate "downstream" processes such as manufacture, assembly, testing, service, and life-cycle support. Guidelines for performing coordinated design are as follows.

Develop a Comprehensive Problem Statement: To ensure optimal chunking of the design, it is important to fully understand long-term market, technological, manufacturing, and life-cycle support needs. How will market needs and customer needs change over time? What aspects of the design will eventually need to be redesigned to accommodate changing technological capability? What upgrades, add-ons, and/or adaptations will be needed? What aspects of the design can or should be standardized? What aspects of multiple versions of the design are common? What aspects of multiple versions of the design are unique? What are the manufacturing, assembly, testing, and service issues and challenges? The problem statement is a clear and concise statement of what the coordinated design is to achieve with respect to "downstream" processes and long-term business needs. When possible, it is often useful to formulate the problem statement as a series of design goals together with constraints that must be satisfied. In developing the problem statement, it is important to (1) thoroughly understand the performance, manufacturing, and life-cycle support issues and problems associated with previous and/or existing designs; (2) articulate a design strategy that supports the business and marketing strategies of the firm; and (3) understand the future to the fullest extent possible, especially with regard to technological change and market change. Remember that customers for the coordinated design include the manufacturing floor, system houses and other suppliers, service personnel, and others who are involved with "downstream" processes. In addition to interviews, careful observation of existing production processes and servicing procedures can yield a great deal of insight.

Create a Part Decomposition Schematic: The part decomposition schematic is the starting point for developing an overall part decomposition optimization. The goal is to coordinate the geometric layout with a logical, well-reasoned "chunking" strategy and assembly concept (structure and assembly sequence), while also applying the part decomposition optimization guidelines of Section 4.1 (see page 44). The schematic should facilitate this by representing the physical elements of the design in a way that provides helpful information and guidance regarding spatial relationships, interactions (material flow, force and energy flow, information flow, etc.), and design constraints (e.g., required use of standard components). In most cases, there is a lot of latitude in how the schematic is drawn. For this reason, it is often useful to generate several alternatives to gain multiple perspectives on how the optimization might be achieved (see Illustrative Examples 5.1 and 5.2 on pages 67 and 68, respectively).

Group the Elements into "Chunks": Remember that "chunks" are collections of physical elements that are grouped together to provide explicit advantages. The "chunking" logic that is used is critical to achieving a coordinated design that minimizes friction in the design. Begin by assuming that each element in the part decomposition schematic will be assigned to its own chunk; then, using the problem statement and the following guidelines, group elements to provide desired benefits and advantages. Always use the problem statement as a guide.

- **Locally close force-flow paths:** Elastic deformation can be a major source of undesirable interactions. It also adds cost, weight, and complexity to the design. Grouping elements to locally close force-flow paths minimizes and controls deformation. See the discussion on page 40.

- **Isolate precision:** Information content is reduced when elements requiring precise location or close geometric integration are part of the same chunk. Avoid tolerance stack-up by confining precise dimensions to a single part.

- **Share function:** When a single physical component can implement several functional elements of the design, these elements are best grouped together. *WARNING*: always be on the lookout for undesirable interactions.

- **Leverage vendor capability:** A trusted vendor may have specific capabilities. Group elements to take advantage of such capabilities.

- **Group similar technology:** Group functional elements that utilize the same technology. For example, grouping all functions that are likely to involve electronics might allow the possibility of a single circuit board.

- **Localize change:** Apply the principle of least commitment by isolating elements that are likely to change into their own chunk or chunks.

- **Accommodate variety:** Group elements together in ways that make it possible to upgrade, add-on, and/or adapt the design according to changing customer and market needs. Isolate differentiating elements in one chunk.

- **Avoid manufacturing, assembly, testing, and service problems:** Group elements in ways that simplify and/or avoid production problems and assembly precedence constraints. Group elements to facilitate subsystem testing as well as quick maintenance and/or service.

Visualize a Coordinated Geometric Layout and Assembly Concept: In addition to working out the dimensional and spatial relationships between the chunks, this key activity requires that an assembly structure (support and integrating scheme) and assembly sequence be envisioned. It may also be necessary to select a particular material class and/or manufacturing process. Developing the geometric layout and envisioned assembly concept not only reveals whether the geometric interfaces among the chunks are feasible, but also identifies assembly constraints and requirements early in the design. Optimize by analyzing and redesigning using the optimization guidelines of Section 4.1 (page 44).

Envision a Possible Assembly Process: In addition to the assembly structure and sequence, the envisioned assembly process may include a target number of workstations (or workers), a material handling scheme, assembly system architecture, factory floor layout, and other consideration that are important to the design such as testing and calibration. In general, the more detailed the assembly process, the more useful it is in guiding design decisions. The degree of detail will depend largely on the completeness of the problem statement, the expertise of the team, and the maturity of the design. The minimum requirement for effective envisioning of the assembly process is to have manufacturing engineering strongly represented on the design team. Often, the desired coordination naturally occurs as the result of timely input, guidance, feedback and insight provided by the manufacturing representative. A more proactive approach, which is sometime called "process-driven design", may also be used. In *process-driven design*, the assembly process plan (assembly sequence, target number of workstations, assembly system architecture, factory floor layout, etc.) is first developed and the product is then designed to be assembled according to the proposed plan. Process driven design requires that the company have extensive experience with design and manufacture of the product. For this reason, mature products such as automobiles and appliances are often best suited for the process-driven design approach. A sophisticated manufacturing organizations that is capable of supporting the process-driven design approach is also an important success factor.

Refine and Optimize the Coordinated Design: The design is refined and optimized when friction is reduced to a minimum, i.e., when all downstream needs are understood, all undesirable interactions are avoided, and information content is minimized. Because the coordinated design steps are tightly coupled, iteration is usually necessary to achieve optimality. In essence, iteration implements the "Guided Design Method" discussed on page 18 of the manual. Keep the following in mind when refining and optimizing the coordinated design:

- Experimentation using simple models and mock-ups can answer questions, raise new questions, and provide understanding that cannot be gained in any other way. Use the user-centered design method (Section 2.6, page 28) as a guide.

- The schematic created in Step 2 will not be unique (see illustrative example 5.2, page 68). Similarly, many alternative geometric layout/assembly concept alternatives are possible. Generate and evaluate several alternatives to ensure that the best possible coordinated design solution is identified.

- Some chunks may be very complex systems in their own right. Each of these chunks may have its own scheme by which it is divided into smaller chunks. The coordinated design guidelines can be used at all levels of design.

- Coordinated design requires a team approach. Industrial design, systems houses, equipment vendors, and tool suppliers should always be involved in decisions that concern them.

Illustrative Example 5.6: Redesign a window air conditioner product using the coordinated design guidelines. An exploded view and geometric layout of the current window air conditioner design is shown below. Examination shows that (1) the assembly structure (sheet metal chassis) lacks clarity with respect to the assembly concept, (2) the piece parts appear to be optimized for sheet metal forming processes, but not for ease of assembly, (3) all elements are grouped into one chunk (236 parts), and (4) there appears to be no logic or plan regarding standardization, model differentiation, functional testing, etc. Most importantly, the existing window air conditioner design has been plagued with production problems. Primary among these are refrigerant leaks. As currently manufactured, leaking units are not detected until final testing of the completed assembly. Repair requires extensive disassembly, rework, and testing which is very costly and time consuming. Other manufacturing issues include noisy operation (vibration, buzz, rattle, etc.), marginal cooling performance due to air leaks and other quality problems, and unsatisfactory appearance due to poor fit and finish.

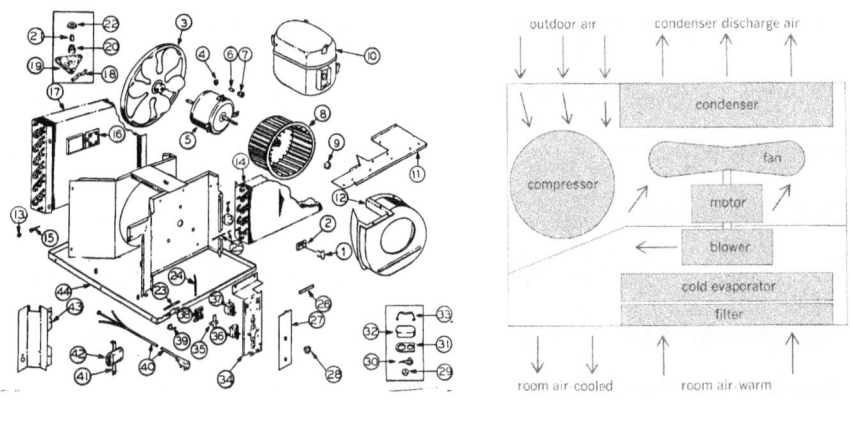

(a) Exploded View   (b) Geometric Layout

Problem Statement: The problem statement is formulated as four design goals.

1. Avoid adding value to defective units.
2. Reduce part count (currently 236 parts).
3. Simplify assembly and reduce assembly content.
4. Provide ability to use different control technologies (mechanical, analog, digital, etc.).

Constraints include the need to use the existing evaporator and condenser heat exchangers, which are internal standard components that are used interchangeably in several different products.

Part Decomposition Schematic: A possible part decomposition schematic is shown below. Since the air conditioner operates on the vapor compression thermodynamic cycle, it should be noted that the compressor, condenser, expansion valve, and evaporator are connected together by copper tubing to form the refrigerant circuit. It is also important to note that a barrier is needed to separate the cold, room (evaporator) side of the air conditioner from the hot, heat rejection (condenser) side.

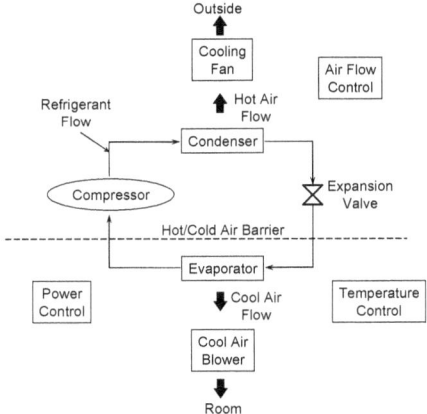

Group into Chunks: Since refrigerant leaks are a major concern, group the compressor, condenser, expansion valve, and evaporator into a "refrigerant system" chunk. To isolate change, group the control functions into a "control module" chunk. Combine the fan motor, hot-air side fan, and cool-air side blower wheel into a "fan system" chunk.

Geometric Layout and Assembly Concept: Based on the part decomposition optimization guidelines presented in Section 4 of the manual, use a z-axis stacked construction approach. As shown below, the assembly concept consists of a base and upper chassis, with the refrigerant system, control module, and fan system located by fixture nests and "sandwiched" in between.

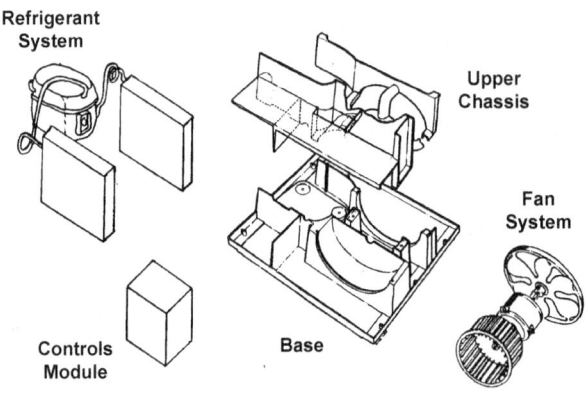

Possible Assembly Process: In accordance with the design goals for leak detection and minimum possible number of assembly workers, an automated assembly process consisting of a serial assembly line with feeder lines is envisioned as shown below. In this process, each chunk is assembled and tested on its own feeder line and then robotically inserted into the final assembly.

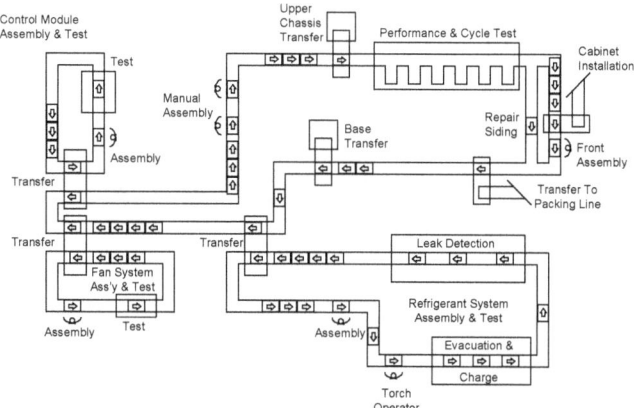

Refine and Optimize: To facilitate the chunking and stacked assembly concept, it is decided to fabricate the base and upper chassis out of sheet molding compound (SMC). This enables a large part count reduction by making it possible to mold nesting and guiding features as well as form the hot/cold barrier and air-flow channels in the base and upper chassis parts. Using SMC for the base and upper chassis, the following iterative improvements are achieved.

Current Design	Initial Redesign	Final Redesign
236 parts	155 parts	120 parts
	Issues: • Air handling • Refrigerant system installation • Too many fasteners, seals, etc. • Complicated tooling • Flexibility and strength of base component	Major improvements: • smaller fan motor • no fasteners in final ass'y • all straight pull molds • hardware commonality • elimination of seals • error proof assembly • unique compressor mounting plate
	26 performance concerns	Process improvements: • assembly sequence • assembly line architecture • coil and tubing connections relocated
	43 assembly issues identified	
	9 tooling and process concerns	

Insights: The initial redesign is refined and optimized by reducing friction (undesirable interactions and information content) to a minimum. Insights gained from the improvement process are summarized as follows.

- All design goals were met or exceeded. Experience has shown that setting "stretch" business, manufacturability and life-cycle support goals as part of coordinated design stimulates innovation and design creativity. Most importantly, by considering these goals early in design, they are almost always achieved, often in unique and unanticipated ways.

- Coordinated design challenges conventional thinking. In addition to improving air flow and cooling performance, the decision to mold the base and upper chassis rather than build them up out of numerous sheet metal components facilitated significant part count reduction and manufacturing simplification. This material substitution decision was not easy however, because it goes against many years of standard industry practice.

- Coordinated design facilitates early identification of undesirable interactions. Consider, for example, the robotic transfer and insertion of the refrigerant system that was originally envisioned. Analysis showed that the concentrated weights of the compressor and heat exchangers connected together by very flexible and "springy" copper tubing made this operation prohibitively difficult. The problem is avoided by a simple assembly process change: combine the refrigerant system feeder line with the final assembly line and use the base component as a fixture for assembling the refrigerant system. To facilitate this change however, the tubing connections had to be relocated away from the SMC base. Doing this was easy because the change was made early in the design. Such a change would probably not be possible after design release or after gaining UL approval.

- By using the Part Decomposition Improvement Method (Section 4.11, page 58), several seals and gaskets were eliminated. These design changes made it possible to eliminate the two manual assembly stations on the final assembly line.

- As part of the design optimization, the base and upper chassis were redesigned to eliminate all camming and side-action in the mold. This not only reduced tooling cost and process cycle time, it also greatly simplified the air-flow channels. As a result, it was possible to use a smaller and less expensive fan motor, thereby producing a significant unanticipated cost savings.

- The material cost for the SMC parts is greater than the cost of an equivalent spot-welded sheet metal fabrication. Because the firm has no experience with large molded plastic parts, making the tough decision to go forward with the SMC design requires faith that the information content reduction will more than offset the cost penalty. Quandaries such as this illustrate the challenges that can sometimes arise when seeking an optimal coordinated design.

5.3 DESIGN BACKWARDS APPROACH

The design backwards approach is a "team based" ***thought experiment*** that is intended to help stimulate the development of a viable coordinated design concept. The exercise is best performed by the whole team including industrial design and outside systems houses and suppliers when their input is important. Begin by envisioning a "coordinated" geometric layout and assembly concept. Using this as a guide, visualize how the physical elements could be designed for ease of assembly and fabrication. Finally, when all constraints are understood, optimize and refine the coordinated design as a system. Repeat the design backwards process making changes as necessary to eliminate problems and to further reduce information content. If anticipated problems and proposed fixes require research, agree to do homework and meet again at a later date (but not too much later). In some cases, the design backwards process may be repeated several times as understanding and design knowledge increases.

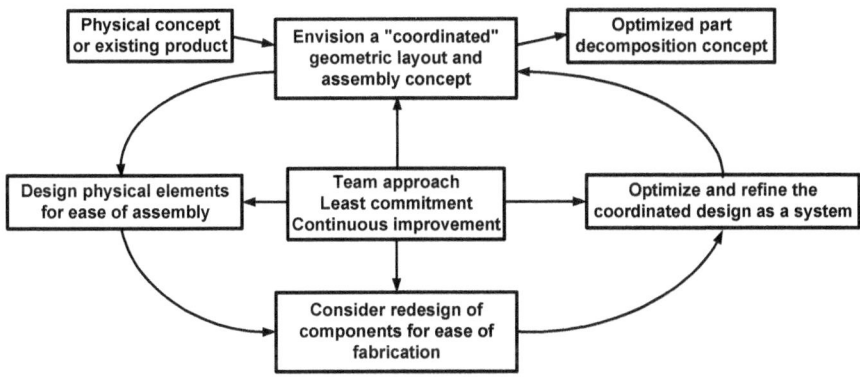

One possible approach for performing the design backwards thought experiment is to use the following two step procedure. To minimize interruptions, it is best for the team to meet "off-site" or in a meeting facility with limited access.

1. Starting with an existing design or proposed new design, use the part decomposition improvement method (page 58) to systematically develop several alternative redesign concepts. A good way to develop a diversified portfolio of ideas is to break into small groups and have each group perform the analysis independently.

2. The whole team then reconvenes and, starting with the redesign ideas, develops alternative assembly process scenarios. These scenarios are then modified and evolved until a coordinated design concept that the team can realistically build-on emerges.

Although the concept developed may not be the one that is eventually used, the design backwards exercise helps to sensitize the team to the opportunities available and creates a shared vision for a workable coordinated design.

5.4 ADVICE, INSIGHTS, AND CAVEATS

Philosophical framework. Coordinated design provides a philosophical framework for developing the part decomposition and should not be viewed as a step-by-step procedure. Rather, coordinated design should be seen as providing the discipline needed to help ensure that all "downstream" needs and constraints are properly identified and considered early in the design process.

Identify downstream process needs early. Early planning of downstream processes such as assembly, packaging, testing, service, recycling, environmental impact, and so forth generate customer needs. These needs help to constrain design decisions that would otherwise be arbitrary. Ambiguity is decreased, information content is reduced, and safety is made inherent.

Consider production needs first. To minimize information content, consider assembly and component fabrication needs prior to optimizing the design from a functional and piece-part cost perspective.

Do not add value to defective assemblies. Envision an assembly process in which defective components and/or assemblies can be identified and eliminated before further value is added.

Group differentiating elements. To help minimize information content, group all elements associated with customization of the design into one or a few chunks, thus limiting customization to as few assembly operations as possible.

Postpone differentiation. To help minimize information content, design so that the production process can be kept the same for all design variants for as long as possible. Ideally, customization should be performed as the last production step.

Plan the wiring layout early. Wire harnesses and other "connective tissue" are often left to the last minute. Fifty percent or more of the wire used in complex electrical and electronic equipment can be eliminated simply by planning the wiring layout before the individual black boxes and subassemblies are designed. Importantly, early planning helps facilitate logical connector location and reduction of randomness by adding features that orient, locate, and restrain loose wires and wire harnesses.

Use low technology on the factory floor. Always design so that the simplest technology possible can be used to fabricate parts and to support assembly. Use the simplest mechanization and automation possible.

Use common manufacturing processes. Design to reduce the number of different processes and operations required. Strive to use the same manufacturing processes everywhere. Make it a rule to avoid special exceptions whenever possible.

Rationalize manufacturing processes. Develop a short list of preferred manufacturing processes and practices and use these in future designs.

SECTION 6

QUALITY BY DESIGN

Each design decision, both large and small, contributes in some way to total quality of the design. Sections 1 through 5 of the manual teach how total quality is maximized by eliminating friction in all its forms and at all levels of the design. In this section, we focus on design considerations that relate specifically to how total quality can be improved by design.

6.1 FAILURE MODE AND EFFECTS ANALYSIS

Failure mode and effect analysis (FMEA) is the name given to a group of activities that are performed to identify possible failure modes of a design, assess the likelihood that a failure might actually occur, and ensure that appropriate corrective action is taken to prevent a failure occurrence. The primary goal of FMEA is to avoid surprises and to prevent unnecessary quality risk from reaching the customer. To identify potential failure modes and their effects, the design team seeks answers to the following key questions:

1. What is the intended function of the component, subsystem, or system?
2. What are the possible failure modes? How could the design conceivably fail to perform its intended function?
3. What would be the effect if the failure did occur?
4. What mechanisms or causes might produce the failure mode?
5. What current controls or counter-measures are provided to prevent the failure or to compensate for it?

Once these questions and others like them have been answered, the risk associated with each potential failure mode is assessed. Typically, three risk factors are considered: severity, occurrence, and detection. Severity (S) is an assessment of the seriousness of the effect that would be produced by the potential failure mode should it occur. Occurrence (O) is an estimation of the probability that a specific mechanism or cause will occur. Detection (D) is an assessment of the ability to detect a potential mechanism or cause or an actual failure before it reaches the customer. Evaluation criteria for each risk factor are based on a point scale.

Design FMEA Evaluation Criteria

Rating Value	Severity (S)	Occurrence (O)	Detection (D)
1	No effect to minor; Defect may be noticed by customer	Rare	Will almost certainly be detected
2	Customer is inconvenienced	Infrequent	Reasonably detectable by current controls
3	Item is operable, but at a reduced level; customer is dissatisfied	Frequent	Detectable before reaching the customer
4	Item is inoperable; loss of function	Very frequent	Detectable only by the customer &/or during service
5	Safety-related catastrophic failure; regulatory noncompliance involved	High to very high	Undetectable until catastrophe occurs

Using the agreed upon rating criteria, the overall risk is computed as,

$$RPN = S \times O \times D$$

where *RPN* stands for "risk priority number". Using a 5 point scale, the *RPN* values will range between "1" and "125". Obviously, an *RPN* value of "125" would be of great concern. On the other hand, and *RPN* value of 1 could probably be completely ignored. *RPN* values between these extremes require interpretation criteria such as that given in the table below. As a general rule, failure modes that have a high severity rating should be given special attention regardless of the resultant *RPN* value.

RPN Interpretation Criteria

RPN Value	Interpretation/Action
$1 \leq RPN \leq 17$	**Minor Risk:** little or no action required.
$18 \leq RPN \leq 63$	**Moderate Risk:** this requires selective design validation and/or redesign to reduce the Risk Priority Number.
$64 \leq RPN \leq 125$	**Major Risk:** high priority. Extensive design revision should be taken to reduce the Risk Priority Number.

The intent of corrective action is to reduce any one or all of the risk factors. Possible corrective actions are typically identified by "brainstorming" or by investigating the actions that were taken for previous or similar designs. When possible, the best corrective action is usually a design change. This is why it is recommended that FMEA type assessments be conducted early in the design process when design changes are easy to make. A design change can reduce any or all three of the risk factors. Also, a design change is the only way that the severity rating and occurrence rating can be reduced. The detection rating, on the other hand, can be reduced by making a design change, or in some cases, by increasing preventative measures such as added inspections on the production line or by use of validation and/or verification testing.

6.2 ROBUST DESIGN

Variability is an enemy of good design. A *robust design* is one that is insensitive to hard-to-control variation. There are typically three sources of hard-to-control variation: the operating environment, deterioration that occurs over time and with use, and product to product variation. Examples include temperature, humidity, input voltage, corrosion, wear, and dimensional variation. Robustness is improved by (1) designing an inherently robust physical concept, (2) optimizing the design parameter settings, and (3) when all else fails, specifying tight tolerances.

Design a robust physical concept: As discussed in Section 1.6, one of the best ways to ensure that the physical concept is inherently robust is to satisfy each functional requirement independently (see Illustrative Example 1.4, page 12). Avoiding undesirable interactions will also improve robustness. The seal design illustrated below is a case in point. In Design A, relative motion between the seal and rotor surface generates frictional heat that causes the rotor to expand. This, in turn, causes the seal to press harder, generate more frictional heat, and wear more quickly. This undesirable interaction is avoided in Design B by designing so that rotor expansion decreases the seal pressure on the rotor. Design B is preferred because it is inherently robust against hard-to-control thermal expansion.

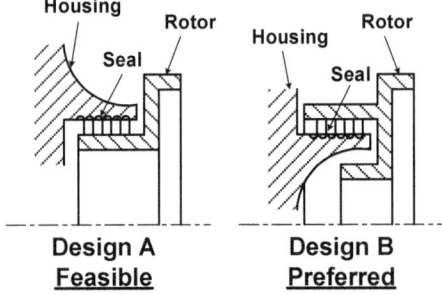

Optimize parameter values: Parameter values are optimized when sensitivity to hard-to-control variation is minimized. As illustrated below, this can sometimes be achieved by taking advantage of non-linear relationships that exist in the design.

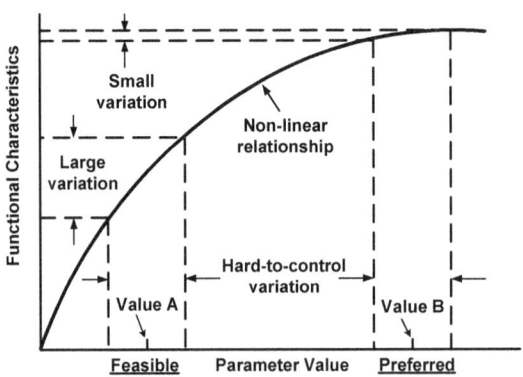

Illustrative Example 6.1: A helical coil spring is to be used in an application where the spring force should be as constant as possible, but the working length of the spring is hard-to-control. Two springs are available: Spring A, which has a high spring rate and Spring B, which has a low spring rate. Which spring is preferred?

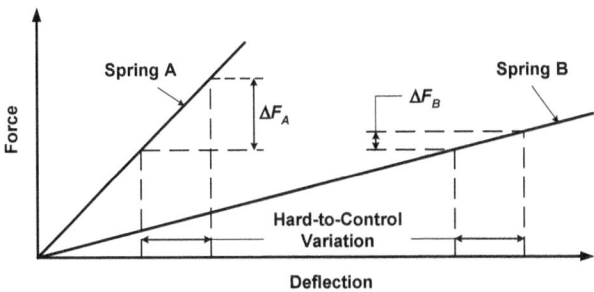

Choice: As illustrated by the spring rate curves, **_Spring B_** is preferred because the force variation (ΔF_B) is smaller for the hard-to-control variation. In general, a low spring rate is less sensitive to hard-to-control variation in working length as well as variation of helical coil spring manufacturing parameters such as free-length and number of turns.

Illustrative Example 6.2: The bearing assembly shown in (a) below wobbles unacceptably due to tolerance stack-up of shaft, needle bearing, and housing dimensions. What minimum bearing length (L_{min}) is required for the percentage of assemblies having unacceptable wobble to be below 1%?

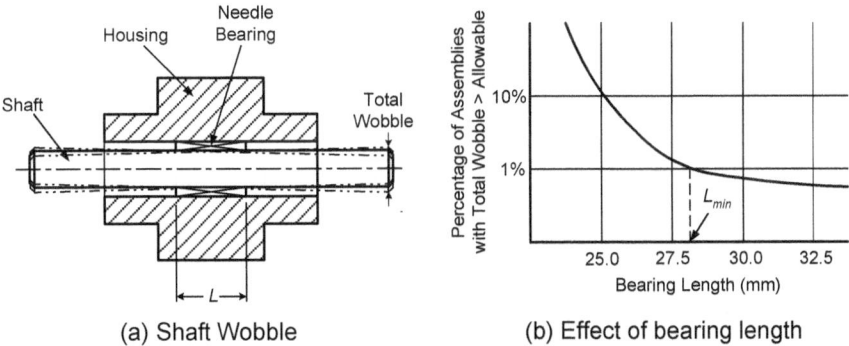

(a) Shaft Wobble   (b) Effect of bearing length

For this simple example, the percentage of assemblies that exceed the allowable wobble is calculated as a function of bearing length using commercially available "Monte Carlo" simulation software. As shown in (b) above, this percentage varies non-linearly with bearing length. In particular, it is seen that the percentage of assemblies that wobble unacceptably will be below 1% when **_L_{min} > 28 mm_**. In situations involving more than one design parameter, identifying a robust value for each parameter is more complicated. For these cases, design of experiment (DOE) techniques such as the Taguchi method are recommended.

6.3 PERCEIVED QUALITY

Perceived quality has to do with the experience the owner/user of the design has with respect to efficiency, maintenance, service, and other perceived consequences and hassles connected with owning and operating the design. In general, perceived quality will be high when friction is low. Perceived quality is further improved by the following practices:

Strive for efficiency: Endeavor to maximize productivity (desired outputs) and minimize waste of all kinds. This is easy to say, but often very hard to do. When the design succeeds, perceived quality is greatly enhanced.

Make the most of materials: Utilize material properties to the fullest extent possible. For example, design so the factor of safety is the same everywhere in a load carrying component or structure (see Section 3). This reduces weight while maximizing performance and capacity.

Make the design durable: *Durability* is the amount of use the design can withstand before it is preferable to replace the design rather than to continue to repair it. Understand customer needs and expectations to achieve the right balance between material cost and purchase price.

Miniaturize: Small products and parts require less material, weigh less, occupy less space, and have smaller footprints making them easier to package, ship, store, display, and carry home. Experience has shown that part elimination, especially by using integral design and by identifying designs that require fewer theoretical parts (see Section 4) is one of the most effective routes to miniaturization.

Maximize serviceability: A design is easy to service and maintain when the part decomposition is optimized as discussed in Section 4. Reduce friction even further by (1) defining a short list of permissible tools and operations early in the design; (2) designing so parts are easily identified, readily visible, and easily reached; (3) providing sufficient clearance for easy removal and reinstallation without the need for excessive disassembly; (4) ensuring that all maintenance points are accessible; and (5) providing self-diagnostics that trip visible warning indicators.

Use common parts: Whenever possible, use common mounts, connections, threads, connectors, fasteners, belts, tubes, hoses, filters, lubricants, etc. Common parts not only simplify repairability, they also maximize availability of spare parts and enable parts scavenging and substitution in emergencies.

Simplify recycleability: (1) Design parts to be easy to rebuild so that they can be reused; (2) design so parts or the scrap from the manufacture of parts can be reused in alternative applications; (3) design parts so that material separation is easy.

Avoid unintended consequences: Anticipate health, safety, and environmental impacts early in the design so that they can be used to constrain design decisions that would otherwise be arbitrary.

6.4 RAM-D

RAM-D stands for reliability, availability, maintainability, and dependability. *Reliability* is the probability that the design will perform its specified function for a specified time under specified operating conditions. *Availability* is defined as the fraction of time a design is able to function. *Maintainability* relates to the average time required to perform preventative maintenance and the average time required to repair a typical failure. *Dependability* is the result of reliability, availability, and maintainability. The design is dependable if it can be relied upon to be available and to perform reliably when needed. Reliability and availability depend on failure rate (the number of failures per unit time). Availability also depends on the mean time to perform a typical repair and on the average time required to perform preventative maintenance. RAM-D is improved when the friction associated with failure rate, repair, and preventative maintenance is reduced. Proven practices for improving RAM-D include the following:

Avoid failure: Design to prevent failures (see Section 3).

Standardize where possible: Use standardized components of known and proven reliability.

Derate purchased components: Operate purchased components such as electronic devices and electric motors at partial full-load capacity. This design practice, which is referred to as "derating", essentially adds to the life of the component by applying an additional factor of safety.

Use redundancy: In parallel redundant designs, multiple duplicate systems are used to ensure continued operation even when one or more systems fail. When cost or weight penalties are excessive, "back-up redundancy" can be used. An automobile spare tire, or better yet, a tire patch kit and air pump are examples.

Isolate wear-out: Incorporate wear-out and failure prone aspects into easily replaceable modules, chunks, or components. Design for easy assembly and disassembly (see Section 4 and 5).

Design in "detectability": This desirable characteristic makes it easy to monitor and/or check for component deterioration and the need for repair or preventative maintenance. Detectability replaces the inconvenience, cost, and time associated with unplanned failure with more predictable planned or preventative maintenance.

Simplify: Minimize information content of the design (see Sections 1 and 4). Simple components and assemblies often have fewer failure modes and less opportunity for error or malfunction. Therefore, identifying the minimum number of simply shaped components that are easily analyzed and manufactured is a highly effective way to improve RAM-D. The idea of simplification extends across all aspects of design.

6.5 ADVICE, INSIGHTS, AND CAVEATS

Design for quality. To design for total quality, the design team must systematically focus on total quality as a design objective. Use the rules of good design to facilitate a systematic approach. The concerns of good design are (1) how well the design satisfies customer needs; (2) how well the product performs over time, during use, and with respect to safety and environmental impact; and (3) how easy the design is to design, manufacture, distribute, sell, service, and support in the field. In other words, what will cause the customer to select or purchase the design? What will delight and satisfy the customer as an owner and user of the design? And, what will make it worthwhile for the firm to design and sell the design? Design for quality by answering these questions.

Focus on the components of total quality. Decompose total quality into "external" and "internal" components as shown below.

External quality affects sales demand and selling price. It is a reflection of how well the design satisfies customer needs. *Internal quality*, on the other hand, relates to the ease of design, manufacture, and life-cycle support and is therefore important to the manufacturing enterprise. Each quality is measured on at least three major dimensions. *Quality of concept* is what makes the design desirable to purchase, *quality of ownership* reflects the experience the customer has in owning and operating the design, and *operational robustness* reflects the designs ability to tolerate hard-to-control variation that influences functionality. *Quality of conformance* measures how well the physical embodiment conforms to design intent, *quality of producibility* relates to the ease with which the design can be manufactured, assembled, inspected and tested, and *manufacturing robustness* gauges the ability of the manufacturing system to tolerate design changes and production quantity (demand) changes that occur due to changing market needs, business needs, and technology innovation. Employ a twofold strategy: (1) determine what factors or design considerations affect each quality component and (2) seek to understand how these factors can be adjusted or influenced *by design* to maximize total quality.

Leverage the benefits of total quality. High total quality not only contributes directly to the firm's bottom line by generating demand and reducing total cost, it also makes manufacture and life-cycle support easier. This, in turn, helps create and foster a more positive company-wide attitude toward quality. The result is continuous company-wide improvement.

SECTION 7

MANUFACTURING ENTERPRISE

When design is carried out within the context of a manufacturing enterprise whose primary purpose is to manufacture and sell products for a profit, the ability to achieve good design is often affected by complex business considerations and by organizational friction generated by the manufacturing enterprise itself. How the manufacturing enterprise enables and/or constrains the design process is therefore of key importance to good design. In this section, we touch on aspects of good design such as standardization, part count reduction strategy, total time reduction, and effective design process that depend, at least in part, on business, organizational, and leadership considerations.

7.1 BENEFITS AND COSTS OF STANDARDIZATION

Standardization is the most far-reaching and all encompassing approach available for reducing information content of a manufacturing enterprise. Standardization works by limiting the number of options and minimizing the information content of the options that remain. Offsetting the benefits of reduced information content are the compromises and implementation costs that accompany most standardization efforts. Finding the best route to standardization therefore requires systematic assessment of the benefits and costs.

Benefits of Standardization: Standardization reduces friction and information content in a variety of ways. When considering standardization, it is important to thoroughly assess all the benefits. Major benefits of standardization include the following.

Setups: A *setup* includes all of the non-value-added activity and material cost associated with changing a manufacturing process or operation from one product to another. Standardization avoids setups by allowing equipment to be dedicated to one product. Friction is reduced because the adjustments, tool changes, material purges, scrap, rework, lost production, lost equipment utilization, and scheduling effort involved in setups is eliminated.

Inventory: Firms hold inventory in the form of *working stock* to compensate for lot size and *safety stock* to hedge against uncertainty. Both forms of inventory are reduced when several nonstandard products are replaced with one standard product.

Material Resource Planning: Specific parts used in each unique product must be stocked in the right quantity by correctly forecasting demand for each unique product. Friction is greatly reduced by standardizing the parts so that any of the unique products can be manufactured using the same standardized parts. Forecasting is improved because the firm needs to only forecast demand for the whole product line, not individual products. In addition, lead time is reduced because the larger quantity of standardized parts required by the whole product line allows more efficient production methods and, since the same parts are used in a variety of different products, there is increased probability that parts will be available when a specific product is ordered.

Design and Development: A product line composed of similar but not identical products can greatly increase friction associated with each individual product because each product may require its own design engineer, set of tooling, specialized maintenance and service procedures, testing and regulatory approval, suppliers, and so forth. In addition to eliminating these duplications, standardization reduces friction because designers become knowledgeable about the strengths, weakness, performance, and other characteristics of the standardized parts. This knowledge facilitates faster and more effective design cycles.

Product Options: To be competitive, many firms must manufacture a full line of products to meet customer needs. To reduce complexity, management is often tempted to reduce the number of options. This will not necessarily reduce friction, however, especially if the same parts are required for the product options that remain. The key to reducing friction associated with product options is to reduce the number of parts by standardizing the parts so that any of the unique products can be manufactured using the same standardized parts (see Section 4.2).

Collateral Benefits: The benefits of standardization ripple in many ways throughout the manufacturing system and supply chain. The whole manufacturing system and supplier network should therefore be considered when analyzing potential benefits.

<u>**Cost of Standardization:**</u> Standardization is not free. If not carefully planned and thought through, it can increase total cost rather than reducing it. It can also alienate customers and cut off access to profitable markets. Some major costs of standardization that must be considered include the following.

Over-Design: Products are "over-designed" when they are better than they need to be, that is, when they are capable of better performance or more functionality than a particular application requires. Over-design is typically accompanied by the following costs:

- Increased weight.
- Increased or extra manufacturing cost.
- Cost penalty increases with production volume. Therefore, over-design is less acceptable for high volume products.
- Performance and efficiency may be reduced producing competitive disadvantage.

To off-set disadvantages of over-design, do the following:

1. Standardize in ways that avoid over-designed parts.
2. Standardize in ways that increase sales demand for the least over-designed product.

Customer Needs: Products proliferate to meet nuances in customer needs. If standardization causes the firm to exit some market niches or causes the customer to prefer a competitor product, revenue may be lost. To avoid this cost, customer needs must be clearly and concisely understood. In some cases, the standardization will be totally unnoticed by the customer while in others, incentive to purchase the "top" of the line may be lost.

Capital Investment: Standardization often requires substantial capital investment in design, tooling, production equipment, employee training, and so forth. For the candidate standardization opportunity to be feasible, the savings produced by standardization must recoup all of these costs. If the capital investment is very high, the payback period may extend over many years. Standardization, therefore, frequently requires managerial vision and commitment at the top.

Standardization Opportunities: Standardization opportunities exist wherever there is excess complexity and friction. "Red flags" that signal the presence of an opportunity include:

- **Proliferation:** products, designed parts, purchased components, etc.
- **Lengthy cycles:** testing, evaluation, verification.
- **Operational problems:** excessive inventory, long manufacturing lead times, inability to satisfy demand, scheduling difficulties, etc.
- **Small lot sizes:** purchased quantities, run quantities, etc.
- **High overhead:** relative to competitors, industry standards, etc.

One way to surface opportunities is by sorting identified sources of complexity and friction into logical groupings that involve some sort of commonality. For example, products that have a similar specification or functionality may logically group together. Parts that are manufactured using similar processes or that perform similar functions, or that have similar geometrical features or shapes could also form logical groupings. The goal is to discover logical groupings that offer realistic standardization opportunities.

The total cost saving potential of an identified standardization opportunity can be assessed by listing its inherent benefits and costs. By evaluating the benefits and costs of a particular opportunity, it is possible to gain immediate insight into the tradeoffs that exist. Also, the process of identifying benefits and costs teaches much about the standardization opportunity. The following table lists benefits and costs for several common standardization schemes.

Scheme	Benefit	Cost
Limit the number of product models (e.g., the Ford Model T)	• No setups • Less design time and cost • Economies of scale • More management focus	• Product may be over-designed & costly • No incentive for repeat buyers to upgrade
Limit the number of product options	• No setups • Less design time and cost • Economies of scale • More management focus	• Product may be over-designed • May loose customers if options not available
Limit the number of feature sizes (e.g., limit number of hole sizes)	• Fewer setups • Fewer tools	• May need to redesign some products • May have to outsource non-standard sizes
Use existing high volume parts in new, low volume products	• Save design time • Save tooling cost	• Part may be over-designed • May require extra parts to adapt to new product

Scheme	Benefit	Cost
Postpone product differentiation (e.g., customize at regional distribution center to meet regional needs rather than building many different models at one location)	• Inventory reduction • Decreased shipping cost and lower tariffs by adding value locally • Better customer service • Increased sales	• Basic product may be more expensive due to added features that allow field customization • Distribution centers need personnel, expertise, and equipment
Limit the types of processes and equipment used (e.g., use only spot welding and buy equipment from one vendor)	• Reduce training and skills required • Reduce spare parts inventory • Reap benefits of being a preferred customer	• Lost capability may add cost and/or compromise functionality • Can't meet some needs
Rationalize purchased components and use in new product designs (e.g., reduce the number of different standard screws used)	• Less design time • Tested and proven parts and processes • Established supplier base, favorable price • Interchangeability between products	Some components may be better and therefore more expensive than needed
Standardize manufacturing processes (e.g., standardize the die shut height of large stamping presses)	• Less setup time • Faster setup may lead to reduced inventory and improved JIT performance	• Cost required to change existing tools and equipment to conform with the standard.

7.2 USE "PREVIOUSLY DESIGNED" PARTS

Instead of designing a new part from scratch, time and effort can often be saved by modifying an existing part design. When this is possible, design of a new part becomes a simple matter of modifying an existing solid model or CAD drawing by changing dimensions and deleting or adding features as required. With the availability of parametric solid modeling software, the process can be as easy as changing values in a parameter table. In addition to saving design time, using previously designed parts to design new parts produces the following benefits:

- Gradual evolution of "repeat parts".
- Gradual standardization of part families.
- Improved cost estimation.
- Improved ease of manufacture.

General Methodology: One possible approach for using previously designed parts to design new parts is as follows:

1. Sort all existing parts into logical part families.
2. Create a master part for each family.
3. Create a CAD library of master parts.
4. Select a master part and modify as required to design a new part.

Many variations on this methodology are possible. The ultimate goal is to reduce friction associated with design and proliferation of new parts.

Part Families: A part family may be defined as a group of related parts that have some specific sameness or similarities. The family of parts concept provides the information necessary to design individual parts as incremental or modular parts. It also provides the basis for rationalizing process planning and forming machine groups or cells that process specific part families. Geometry based part families have similar features such as geometric shape. Process based part families share similar processing steps and requirements.

Illustrative Example 7.1: A manufacturing company has identified the family of rotational parts shown below. Create a "master part" for the family.

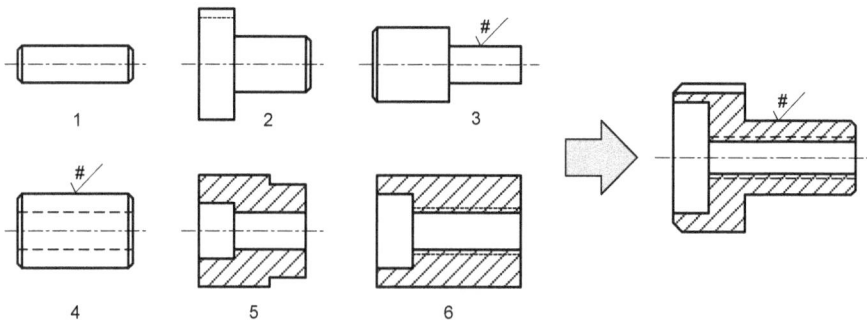

Family of Rotational Parts   Master Part

Repeat Parts: Using "previously designed" parts frequently facilitates the discovery of *repeat parts*, that is, parts that can be used for similar or different purposes in different designs. A mounting plate or bracket, for example that is used to mount a particular component in one design becomes a repeat part if it can also be used to mount the same or a different component in another design. Similarly, a spacer might also serve as an axle, or lever, or stand-off somewhere else. Repeat parts are usually "discovered" rather than intentionally designed because it is difficult to imagine how the same part can be used in different designs and in dissimilar applications.

Part Standardization: Using "previously designed" parts is likely to result in the gradual standardization of part families. By starting with a master part, designers will naturally tend to simplify the design task by using the same fillet radiuses and other detail features in new designs. Over time this practice will result in fewer variations and in more standardized part family features. In some situations, it may be possible to accelerate this process, or even better, to optimize the results by implementing a long-range standardization plan. Goals of such a plan should include the following:

- Minimize the number of different part families.
- Minimize the number of parts within each family.
- Minimize the number of different individual features used in each part.

Cost Reduction: Using "previously designed" parts not only reduces design cost, it also reduces manufacturing costs by facilitating economies of scope and scale. Because the part is based on a master part, the process plan for making the master part can be used to make the designed part simply by leaving out the steps that are not needed. Also, the same group of manufacturing processes or manufacturing cell can be used to produce the part thereby greatly simplifying scheduling and other manufacturing complexities. Finally, because the new part is similar to all other parts in the family, it increases manufacturing volume which justifies automation and other cost and time saving investments.

Group Technology: This is a powerful approach for reducing the information content of the manufacturing system. Group technology (GT) seeks to exploit the sameness or similarity of manufactured parts based on their geometrical shape and/or similarities in their production process. Traditionally implemented by utilizing classification schemes to identify and understand part similarities, new concepts for generative coding and integration with solid modeling data bases is making GT more cost effective. In addition to facilitating the use of previously designed parts, GT allows manufacturing engineers to decide on more efficient ways to increase system flexibility by streamlining material flow, reducing setup time and floor space requirements, and standardizing procedures for batch processes. Standardized process planning, accurate cost estimation, efficient purchasing, and assessment of the impact of material costs are among the benefits GT promises. The competitive edge offered by GT when teamed with the company's computer is significant and definitely bears serious consideration.

7.3 BEST PRACTICES FOR EARLY DESIGN DECISIONS

Decisions made during the early stages of design are especially critical because of the large impact they have on good design. The following best practices for improving the quality of early design decisions have proven to be key success factors for good design.

Use a Disciplined Design Process: A disciplined design process is essential for good design. Key elements of a disciplined approach include the team approach, formal design reviews, and best practice guidelines. The team approach makes needed design knowledge available to support each critical early design decision. Formal design reviews provide an independent check to insure that all customers needs have been considered and that the design will yield a satisfactory profit. Best practice guidelines provide the guidance, checklists, structured methods, and tools needed to ensure a systematic and disciplined design approach.

Develop a Design Strategy: A clearly articulated design strategy helps narrow choices, makes the best choices more obvious, and constrains arbitrary choices. A design strategy can be based on many different considerations. For example, as discussed in Section 5, a process driven design strategy would cause the product to be designed in a way that allows a particular assembly sequence or method of manufacture to be used. A product family strategy, on the other hand, might cause the product to be designed to allow the use of "building block" components or other standardized modules or "chunks". A well-defined design strategy can be the single most important consideration for improving the quality of early design decisions.

Understand Customer Needs: High quality early design decisions depend on comprehensive understanding of both external and internal customer needs. *External* customer needs include the needs of those who distribute, buy, use, and service the product. *Internal* customer needs relate to business needs and strategies, manufacturing needs, and life-cycle support needs. For some design situations, understanding the problem of design requires identifying external customer needs and developing an appropriate design specification. In other situations, the internal manufacturing needs of a product family or the manufacturing constraints imposed by an existing facility must be understood. Good design requires an all-inclusive understanding and characterization of all needs, both external and internal.

Generate Many Alternative Physical Concepts: The likelihood of identifying and selecting the best physical concept is greatly increased when many different alternative concepts are proposed and systematically considered.

Generate Many Alternative Part Decompositions: The likelihood of identifying and selecting the most appropriate part decomposition for ease of assembly, ease of component manufacture, ease of service, and high total quality is greatly increased when several alternative part decompositions are considered.

Use Disciplined Evaluation: The selection of a particular physical concept and part decomposition defines the design and irreversibly establishes many characteristics of the design including total cost, time and quality. The use of an explicit and disciplined selection process helps insure objectivity and customer focus. It also helps guide the team through the decision making process, documents the process, and often provides additional and/or unexpected insights into ways to further improve the design.

Avoid Undesirable Interactions: The presence of an undesirable interaction can make good design virtually impossible. Undesirable interactions are almost always the result of poor early design decisions. It is therefore essential that the team be especially attentive to undesirable interactions during the early stages of the design. One of the best ways to avoid undesirable interactions is to use the "guided design method" (page 18) and the logical building block method" (page 13) discussed in Section 1.

"Just Build It": The best and quickest way to verify an idea is to build a simple physical model or mock-up. Simple models quickly resolve design uncertainties and provide insight and understanding that is difficult to obtain in any other way. Early experimentation with hardware embodiments of ideas and concepts raise "unasked" questions and identify latent needs. It can also provide insight into part decomposition approaches and consequences. The "just build it" experimental approach to early design decision making counters the tendency of the modern engineer to resolve uncertainty using engineering analysis. Analytical approaches can be time consuming and often require more detail design information than is typically available during conceptual design. The "just build it" philosophy avoids the need for detail design information, saves valuable time by resolving uncertainty quickly and unambiguously, and provides insights that can't be obtained in any other way.

Benchmark Competitor Designs: Long-term market success requires that the design be superior to the competition with respect to all critical considerations such as functionality, performance, capacity, cost, and quality. Benchmarking the competition and "reverse engineering" best in class products provides the design knowledge and insights needed to make design decisions that lead to superiority.

Utilize a "What-if" Mentality: The time to question design decisions and to explore all possibilities and options is in the early conceptual stages of design when maneuvering space is wide and hardware is remote. A "what-if" mentality constantly strives to find a better way, seeks to minimize friction and downstream consequences, and maintains openness to rethinking and reconsidering all decisions.

7.4 STRATEGIES FOR REDUCING TOTAL TIME

Re-Engineer the Design Process: Eliminate friction in the design process. Make organizational and procedural changes that enhance communication, encourage design for manufacture, and simplify workflow. Use concurrent engineering practices and the team approach. Co-locate design and manufacturing engineering. Use technologies such as teleconferencing, e-mail, and the Internet to overcome wide geographical separations. Carefully define the product realization process so that the design team knows exactly what it must do during each stage of the process. Institute design reviews to ensure economic viability of the design and to facilitate simultaneous achievement of product manufacturability and tight schedule commitments.

Use Integrated Design Systems: Eliminate friction in design systems. Integrate the CAD/CAM/CAE system so that design and manufacturing engineers work on compatible systems that share information in a seamless environment. Employ computer-aided design tools and methods as an integral part of the design realization process. Utilize solid modeling and parametric and feature based CAD/CAM tools. Computerize, automate, and integrate engineering analysis, manufacturing, and inspection whenever and wherever possible.

Utilize Computational Design: Many product development processes rely on experimental or "test and fix" methods involving costly and time consuming construction and testing of one or more physical prototypes or test beds. Eliminate these practices and the friction associated with them by developing a "science base" that allows the physical testing to be replaced with computer simulations and other analytical techniques.

Standardize where Possible: Reduce design time by leveraging commonalities present in different product models and variants. Do this by developing product architectures that facilitate the use of modular and standardized subsystems and building block parts. Rationalize the variety of choices available for purchased components such as threaded fasteners and ball bearing. New products are then quickly designed since only the unique, non-standard aspects of the new product must be designed. Time required for component testing, vendor selection and qualification, and so forth is also reduced.

Streamline Order Processing: Use the computer to take time and friction out of the order fulfillment process. Develop computer software that enables a sales representative to negotiate and complete a purchase order at the customer's place of business. "Chunk" the design to facilitate customization or design so that customization is performed at the end of the production line. Develop software that processes the specific customer requirements and returns a set of approval documents for immediate review and acceptance by the customer. The best solutions to this type of order engineering scenario are generally realized when the design is developed with a particular order engineering approach in mind. Such designs illustrate how a well conceived design strategy can create significant competitive advantage for the firm.

Reduce Manufacturing Lead Time: *Manufacturing lead time* (MLT) is the time required to process the product through the plant. A short MLT requires short process cycles, short non-operation times (handling, storage, inspection, etc.), and minimal friction and waste at all levels. Widely used strategies for reducing MLT are listed below. When coordinated with the design, these strategies yield synergistic benefits that often exceed expectation (see Section 5).

- **Specialize:** Use special purpose equipment designed to perform a particular operation very efficiently.

- **Combine Operations:** Reduce the number of distinct production machines or workstations through which the part must be routed by performing more than one operation at a given machine. Setup time, material handling effort, non-operation time, and floor space requirements are reduced.

- **Perform Operations Simultaneously:** Perform two or more processing (or assembly) operations simultaneously.

- **Integrate Operations:** Link several workstations into a single integrated mechanism using automated work handling devices to transfer parts between stations. In effect, this reduces the number of separate machines through which the product must be scheduled. With more than one workstation, several parts can be processed simultaneously, thereby further increasing overall throughput.

- **Increase Manufacturing Flexibility:** Maximize utilization of production equipment and reduce cycle time by using programmable automation and flexible automation. Programmable automation places the sequence of operations under computer control to accommodate different product configurations. CNC lathes and CNC milling machines are examples. Flexible automation facilitates the production of a variety of products (or parts) with virtually no time lost for changeovers from one product to the next. Flexible machining centers are an example (see Illustrative Example 4.2, page 48).

- **Automate Material Handling and Storage:** Automated material handling and storage systems eliminate large amounts of non-operation time.

- **Perform On-Line Inspection:** Incorporate inspection into the manufacturing process thereby reducing scrap, improving manufactured quality, and avoiding time wasted making and inspecting poor-quality product.

- **Optimize Process Control:** Reduce cycle time and improve manufactured quality by optimizing control schemes used to operate individual processes and associated equipment.

- **Optimize Plant Operations Control:** Optimize control at the plant level to manage and coordinate the aggregate operations in the plant more efficiently.

- **Utilize Computer Integrated Manufacturing (CIM):** Integrate engineering design and other business functions with factory operations through extensive use of computer applications, computer data bases, and computer networking.

7.5 CREATE A VISION FOR GOOD DESIGN

Avoid inequality of disciplines: High quality broad band-width communication is essential for good design. In some companies, engineering is king, while in others, marketing or manufacturing or some other discipline controls. It is important to recognize inequalities and to balance these with processes and structured methods that promote good communications between all functions and at all levels.

Clarify protocol: Who is the boss? Who has final say? How is team behavior to be rewarded? Who controls the budget? Who sets the schedule? How should suppliers be selected and involved in the early stages of the project? Questions such as these must be clearly and unambiguously answered.

Break down walls: Hard to change evolutionary designs tend to engender functional fixedness and organizational "silos" that isolate functional groups and block effective communication and interaction. Often, structured methods such as the "part decomposition improvement method" (Section 4, page 58) can help break down these walls.

Eliminate negative cultural attitudes: Effective communication and interaction depend on openness and a willingness to listen. In some organizations, underlying cultural attitudes can become serious obstacles to good design. The following examples typify attitudes that signal cultural problems:

- "we've always done it this way"
- "give me my targets and let me do my thing"
- "you don't understand the problem"
- "we don't have time for this"
- "we're the market leader, we must be doing something right"

Establish standards: Good design requires a company-wide focus on the excellence of the design, not on who is responsible for what. Often, this focus is provided by a high-level directive that motivates all employees and puts in place a supportive organizational framework. High-level directives are most effective when they convincingly communicate basic beliefs such as "excellence", "to be the best", or simply "quality". "ISO 9000" certification and "Six-Sigma Quality" are examples of standards that motivate good design.

Be committed: Good design requires management commitment to methods and practices that foster good communication and employee interaction, with continued encouragement, nurturing, and reinforcement.

Co-locate disciplines: Good design becomes more possible with close physical proximity of design, manufacturing, marketing, purchasing, and other key disciplines within the organization.

Provide leadership: Ultimately, achievement of good design depends on managers and executives who exhibit cooperation and model the proper attitude and belief that good design is achievable and desirable.

APPENDIX A: *DESIGN REVIEW CHECKLIST*

Use the following questions as a general guide for both preparing for and conducting design reviews.

☐ Have customer needs been properly identified and understood? Is the functional specification in harmony with customer needs? Is the team satisfied that the right problem is being solved?

☐ Has the design been verified by direct feedback from external and internal customers? Is the team reasonably sure that there will be no surprises, either in the marketplace, or in the field, or on the manufacturing floor?

☐ Are all causes and effects clearly understood, predictable, and unambiguous? Is the team reasonably sure that there will be no hard to explain performance or manufacturing problems?

☐ Is the design inherently safe? Have all failure modes been identified, assessed for severity, and appropriately guarded against by design? Are users, bystanders, and the environment protected from harm at all times?

☐ Are the undesirable effects of structural and component deformation associated with each force-flow path clearly understood? Is the design compatible with these effects, i.e., have potential undesirable interactions been identified and avoided?

☐ Have the benefits and costs of standardization been evaluated? Has an appropriate standardization strategy or plan been implemented?

☐ Has all potential and/or relevant market, technological, and manufacturing change been identified and evaluated? Has appropriate provision for change been provided in the design?

☐ Is the assembly sequence and assembly structure understandable, unambiguous and well-thought out? Has the assembly work content been minimized using the principles of Section 4? Are precedence constraints understood and minimized? Does the design facilitate testing during manufacture so that no value will be added to defective products?

☐ Is the design easy to manufacture and assemble? Are all processing steps predictable and unambiguous? Has the design been analyzed using the Part Decomposition Improvement Method (page 58)? Is the team satisfied that information content has been reduced to a minimum?

☐ Is the design easy to test, service, and maintain? Has the need for special skills and tools been avoided or minimized? Can wear and deterioration be easily detected and/or monitored?

APPENDIX B: *KEY BEST PRACTICES*

The following best practices are key to the Design for Everything approach and should be kept in mind at all phases and stages of design.

Observe the production-consumption cycle: To gain knowledge about the use environment and production line, study it. Ask questions, interview users and workers, send out surveys, etc. To gain wisdom and insight, observe it. When you observe, you learn what is really going on.

Always choose from multiple alternatives: The probability of making high quality design decisions increases greatly as the number of alternatives considered increases. This is because generating alternatives forces the team to explore the whole design space. Always resist the tendency to go with the first idea that you think of. Good design results from doing the hard work of generating and selecting from many viable alternatives.

Understand how components deform: In mechanical systems, experience has shown that the underlying cause of almost all hard-to-fix or hard-to-explain manufacturing and/or operational problems involves elastic deformation of components and/or structures. Remember that real materials deform under load. If force and force-flow occur at any stage in the production-consumption cycle, there will be deformation. To avoid undesirable interactions, understand how this deformation affects the design.

Challenge separate parts: More than anything else, separate parts drive total cost. Never blindly accept the need for a separate part.

Plan the assembly sequence early: Planning the assembly sequence forces the team to think about downstream processes. Features that facilitate assembly can be provided, potential production problems can be anticipated and avoided, and the transition into production can be performed more easily and more quickly.

Plan the wiring layout early: In designs that involve electrical interconnection, planning the wiring layout early forces the team to consider integration issues early in the design when they are easy to deal with. Wire length can be minimized, connector location can be optimized, and potential quality risks and production problems can be identified and avoided.

Analyze the design for improvement opportunities: No matter how carefully you have designed for manufacture and assembly, always analyze the design using the Part Decomposition Improvement Method (page 58) at least once before release. Almost all designs can be improved using this method.

Standardize and rationalize everything: Limiting choices saves design time, leverages experience, and reduces total cost. Purchased parts are a good place to start, but all aspects of the design that involve multiple choices are targets. Remember that benefits accrue over time, so staying the course is essential.

LEARN MORE

The references cited in The Design for Everything Manual are listed below. References [1] and [2] offer a good starting point for those wishing to delve deeper into the rules of design and principle based design. Reference [3] is an easily understood primer on common sense basics of mechanical design that provides straight forward coverage of topics such as exact constraint design, force-flow planning, and so forth. In addition to addressing all aspects of the product design and development process, reference [4] includes relevant discussions regarding the gathering of customer needs, problem decomposition, systematic creativity, and platform design.

References:

1. G. Pahl and W. Beitz, *Engineering Design: A Systematic Approach*, K Wallace (ed), The Design Council, London, 1988.

2. N.P. Suh, *Axiomatic Design*, Oxford University Press, New York, 2001.

3. J. G. Skakoon, *The Elements of Mechanical Design*, ASME Press, New York, 2008.

4. K.T. Ulrich and S.D. Eppinger, *Product Design and Development*, McGraw-Hill, New York, 1995.

ABOUT THE AUTHOR

Henry W. Stoll is professor emeritus of mechanical engineering at Northwestern University, Evanston, IL. He has extensive industrial experience in mechanical engineering design and consults widely with industry. Before retiring from full-time teaching in 2008, Professor Stoll taught product design, mechanical systems design, and manufacturing for many years. Dr. Stoll is the author of two books and several book chapters on design for manufacture and structured design methods. He has presented more than 100 on-site design for manufacture, quality by design, and design for everything workshops to industry.

www.ingramcontent.com/pod-product-compliance
Lightning Source LLC
Chambersburg PA
CBHW051332170526
45166CB00002B/780